# CHINA STATISTICAL ABSTRACT 1988

THE CHINA STATISTICS SERIES

**China Statistical Abstract 1988**

**China Social Statistics 1986**

The China Statistics Series publishes revised English language editions of China Statistics prepared by the State Statistical Bureau of the People's Republic of China (SSB). The English Language Editorial Board, composed of representatives of the China Statistical Information and Consultancy Service Center, an affiliate of the SSB, and the China Statistics Archives, a research center of the University of Illinois at Chicago, is responsible for the translation and revision of these volumes.

# CHINA STATISTICAL ABSTRACT 1988

William T. Liu, *Series Editor*

THE CHINA STATISTICS SERIES

PRAEGER

New York
Westport, Connecticut
London

ISBN: 0-275-93214-1
ISSN: 1040-7979

First published in 1989

Praeger Publishers, One Madison Avenue, New York, NY 10010
A division of Greenwood Press, Inc.

Printed in the United States of America

The paper used in this book complies with the
Permanent Paper Standard issued by the National
Information Standards Organization (Z39.48-1984).

10 9 8 7 6 5 4 3 2 1

# UIC

The University of Illinois at Chicago

China Statistics Archives
Suite 700 South
1033 West Van Buren Street
Chicago, Illinois 60607-9940

# CSICSC

China Statistical Information and
Consultancy Service Center

38 Yuetan Nanjie, Sanlihe,
Beijing, Peoples Republic of China

# TABLE OF CONTENTS

*Section IV*
## Industry



*Section V*
## Transportation, Postal and Telecommunication Services

*Section VI*

## Investment in Fixed Assets

*Section VII*
## Finance, Trade and Price

*Appendix*

# PREFACE AND ACKNOWLEDGEMENTS

**China Statistical Abstract 1988** was completed to offer essential basic information on China's economic and social development in 1987, and to illustrate the major trends in this development since 1978. This book contains key statistics for principal sectors of the national economy by selected years since 1949 and by each single year after 1978. There are nine sections: General Survey; Population and Labor Force; Agriculture; Industry; Transportation, Postal and Telecommunication Services; Investment in Fixed Assets; Finance, Trade and Price; Peoples Livelihood; Education, Science, Culture, Public Health and Sports. **"Statistical Communique of the Statistical Bureau of the People's Republic of China on China's 1987 Economic and Social Development"** and a glossary of major indicators are included at the end of the book.
With the exception of the total surface area of the country, figures for Taiwan are not included in the national figures. Some figures in this survey are provisional, and the final ones will appear in the **Statistical Yearbook of China.**

**Statistical Bureau of the People's Republic of China**

**China Statistical Abstract 1988** is a revised English language edition of the official China Statistical Abstract published by the State Statistical Bureau of the People's Republic of China (SSB). The SSB compiled the statistics contained in this volume and prepared the "Statistical Communique". The English Language Editorial Board was responsible for the translation and annotation of this volume. The English Language Editorial Board is composed of representatives of the China Statistical Information and Consultancy Service Center, an affiliate of the SSB, and the China Statistics Archives, a research center of the University of Illinois at Chicago.

This is the first volume of a series of statistical annuals which will be published by the Praeger Publishers. Future volumes will include the: **China Social Statistics 1986, China Urban Yearbook 1988, China Rural Yearbook 1988, China Industrial Yearbook 1988,** and the **China Price Yearbook 1988**. The China Statistics Archives also publishes **China Statistics Monthly**, which is also available on a quarterly diskette.

The editorial board would like to thank Professors Carmel Chiswick, Mo-yin Tam and Paul Peiper of the Economics Department at the University of Illinois at Chicago for their assistance.

**English Language Editorial Board**

# EXPLANATORY NOTES



This glossary is provided to assist the reader who is unfamiliar with Chinese statistics. The statistical system of China was originally designed to serve the needs of government economic planners and data are frequently divided into categories that reflect administrative needs. More recently, the collection and reporting of statistical data in China has undergone adjustments in response to the reform of China's economy. These reforms are ongoing and will undoubtedly lead to additional changes in the near future.
The concepts and terms explained below are provided to clarify certain basic terms. More detailed information on definitions and concepts can be found in the China Statistics Archives Working Papers Series published by the University of Illinois at Chicago.

## *Basic Terms*

### Ownership

1. State owned units
   *(including government agencies, also refered to as units owned by the whole people)*
2. Collective owned units
3. Individual owned units
   *(refer to the units owned by one or more Chinese individuals)*
4. Joint ownership units:
   *(a) State and collectives*
   *(b) State and individuals*
   *(c) Collective and individuals*
   *(d) Chinese and foreign investors*
5. Foreign owned units owned by:
   *Overseas Chinese or Chinese businessmen from Hong Kong and Macao*
   *Foreign investors*

China's statistical data are often broken down by type of owership. For statistical reporting, the categories 4 and 5 are generally combined into one category called Other Ownership units.

### Material Production vs Non-material Production

1. Material production sectors include five sectors:
   *(a) Agriculture*
   *(b) Industry*
   *(c) Transportation*
   *(d) Construction*
   *(e) Commerce*

2. Non-material production sectors include all kinds of services other than commerce.

Commerce is statistically classified as a material production sector.

### Productive vs Non-productive Investment

An investment project is classified as productive or non-productive according to whether it is directly related to material production.

1. Productive projects refer to that increase material production or that directly serve material production.

2. Non-productive projects refer to residential buildings, public services and facilities for scientific research , culture, education, and health care.

**Industry**
Includes:
1. Extraction of natural resources, including mining and lumbering.
2. Processing of agricultural products.
3. Manufacture of industrial products.
4. Repair of capital goods.
5. Electricity generation and supply, water purification, and gas production.

**Society**
Indicators may be referred to as "of society" or "social", e.g., total investment of society, social labor force. The addition of the words "social" or "of society" means that data include all forms of ownership. For example, "total investment" without other specification generally refers to state owned units only, while "total investment of society" refers to state owned, collective owned and other ownership units. "Social labor force" includes all individuals employed and self-employed by units of all kinds of ownership in urban and rural areas.

**Prices**
There are mainly three kinds of prices:
1. List prices (plan prices), which are determined by government agencies.
2. Negotiated prices, which state and collective owned units can adjust within limits, usually upper limits, set by government agencies.
3. Free market prices, which are determined by market supply and demand (government agencies may also set upper limits for free market prices).

## *Product and Income Statistics*

**Gross Output Value**
The sum of output value (list price x quantity produced) of each enterprise in a given sector.
Note that in summing up the output value of enterprises to get Gross Output Value of a sector, the value of intermediate goods is double counted.
The sum of Gross Output Value of Agriculture, Industry, Transportation, Construction, and Commerce, the five material production sectors of the economy, is defined as Total Output Value of Society.

**Breakdowns of Gross Output Value of Industry**
1. Three categories of ownership are commonly used:
   *State owned industry*
   *Collective owned industry*
   *Other ownership industry*

Additional categories, such as individual industry in urban and rural areas, township and/or village run industry, are also used. However, adding these categories can create overlapping. For example, township and village run industry involves several kinds of ownership, collective, joint, and individual ownership.

2. Light and Heavy Industry
   *Light industry produces consumer goods and handtools.*
   *Heavy industry produces capital goods.*

**Gross Output Value of Industry in Rural Areas**
From 1958 to 1984, enterprises run by townships, formerly people's communes, are included in Gross Output Value of Industry. Enterprises run by villages, formerly production brigades and teams, are included in Gross Output Value of Agriculture. Since 1984, village run industries were transferred from agriculture to industry. Township and village run industries are often combined into one category. However, this category includes several kinds of ownership, collective, joint, and individual ownership.
In rural areas, for cotton ginning, rice husking, wheat milling, animal slaughtering, and sewing, only processing charges instead of the whole value of products are counted in Gross Output Value of Industry.

**Net Output Value**
The sum of value-added of each enterprise in a given sector. Note that Net Output Value does not have double counted value.

**National Income**
The sum of Net Output Value of Agriculture, Industry, Transportation, Construction, and Commerce, the five material production sectors of the economy.
Note the coverage of National Income in Chinese statistics excludes the value added in "non-material production sectors." China's National Income is approximately equivalent to the United Nations' Net Material Product (NMP) .

**Gross National Product**
The sum of value-added and depreciation for all sectors, "material production " and "non-material production", of the economy. Gross National Product has only recently been used in China.

### Classifications

The employment statistics can be classified by type of ownership and by urban and rural locations.

Eight categories are distinguished:

1. Employees of state owned units (including government agencies) in urban areas.
2. Employees of state owned units (includin government agencies) in rural areas.
3. Employees of collective owned units in urban areas.
4. Employees of collective owned units in rural areas.
5. Individual laborers in urban areas.
6. Individual laborers in rural areas.

Individual laborers include individuals who own businesses, usually small scale enterprises, and their employees.

7. Employees of other ownership units in urban areas
8. Employees of other ownership units in rural areas.

### Social Labor Force

Refers to all individuals employed or self-employed in urban and rural areas, including all the eight categories in the above classifications.

### Staff and Workers

This is a literal translation of the Chinese term *zhigong*, which includes categories 1, 2, 3, and 7 in the above classifications.

### Collective and Individual Laborers in Rural Areas

Includes categories 4,6, and 8 in the above classifications.

The coverage can be otherwise specified as:

1. Persons working under the responsibility system in collective units, the majority of whom are farmers.
2. Individuals employed by township and village run industries.
3. Individuals employed by units jointly owned by state and collectives, state and individuals, collective and individuals, Chinese and foreign investors, or units wholly owned by Overseas Chinese, Chinese businessmen from Hong Kong or Macao, or foreign investors, in rural areas.
4. Individual laborers, individuals who own businesses and employees of these businesses.

### Number of Persons Newly Employed in Urban Areas

Refers to all persons having registered with the labor bureau and have found jobs, and secondary school graduates who have registered with their street committees and found jobs, during a given period.



### Investment in Fixed Assets

It includes two major components, capital construction and technical updating and transformation. Sometimes a third category, other investment in fixed assets, is used to refer to investment not covered by the first two categories.

In principle, capital construction covers new construction and expansion projects, whereas technical updating and transformation covers renewing, replacing and rebuilding of existing fixed assets.

Investment in Fixed Assets refers to state owned units only, unless otherwise specified as Total Investment in Fixed Assets of Society, which also includes investment by collective and individual owned units.

### Capital Construction Investment

Investment in new projects, including construction of a completely new facility, or an addition to an existing facility. It includes construction of plants, mines, railways, bridges, harbors, water conservation facilities, stores, residential facilities, schools, hospitals, and purchase of machinery and equipment, vehicles, ships, and planes.

### Technical Updating and Transformation

Investment in projects to renew, modernize or replace existing assets.

### Breakdowns of Capital Construction and Technical Updating and Transformation

1. Source of funds:
   *(a) Central government budget funds are included in central plan, and allocated from the central government budget.*
   *(b) Self-raised funds are from all other sources, including appropriations made by local government agencies at all levels.*

2. Use of funds:
   *(a) Building and installation work includes the cost of construction (excluding management cost of construction units), and equipment installation.*
   *(b) Purchase of equipment includes the cost of purchased or self-made equipment, tools and instruments that accord with the standards of fixed assets.*

*(c) Others include management cost of construction units, surveying and designing costs, land purchase, and other miscellaneous costs not covered in the above two categories.*

**3.** Size of projects:
*The size of a project is decided by the designed production capacity or the total value of investment, specified in the "Standards for Large, Medium and Small Projects" promulgated by the government. This regulation was revised five times, in 1958, 1962, 1972, 1977, and 1979. Therefore, the figures given for projects of different sizes in different years are not comparable.*

**4.** Sector:
*The investment projects are classified into different sectors according to their major products or the main purposes of the projects, irrespective of administrative agencies to which they belong. For example, machine building factories and schools set up by the Ministry of Metallurgical Industry are classified under the categories of machine-building industry and education sector respectively.*

## *Price Indexes*

There are three major price indexes:
***Retail Price Index***
***Cost of Living Index***
***Free Market Price Index***
These indexes differ in the items (goods and services) covered and the prices used in calculating each price index.

### Retail Price Indexes
There are two kinds of retail price indexes:
*Retail Price Index of Society*
*Retail Price Index of State Owned Commerce*
Both cover all consumer goods and the agricultural producer goods sold to rural laborers. Retail Price Indexes exclude services and the agricultural producer goods sold to state owned units. Different kinds of prices are used in calculating retail price indexes.

Retail Price Index of Society, often referred to as Overall Retail Price Index, is calculated using all three kinds of prices, i.e., list prices, negotiated prices, and free market prices. Since different kinds of prices are used, the index is a weighted average with each price weighted by the proportion of total expenditure on the good in question.

Retail Price Index of State Owned Commerce is calculated using list prices only.

Retail price indexes are based on periodical sample survey. The sample set in 1984 includes 300 goods in more than 180 cities and 330 goods in more than 190 county towns.

### Cost of Living Indexes
There are three kinds of Cost of Living Indexes:
*Cost of Living Index of Staff and Workers,*
*Cost of Living Index of Rural Laborers*
*Comprehensive Cost of Living Index*
*(which covers both urban and rural areas)*

Unless otherwise specified, the Cost of Living Index refers to that ofStaff and Workers only.
Cost of Living Indexes cover consumer goods and services, which are considered necessities for daily life as determined by the State Statistical Bureau. The coverage of goods is more limited than Retail Price Indexes, but Cost of Living Indexes include services, while Retail Price Indexes do not.
Currently 325 goods and services are selected for calculating the Cost of Living Index of Staff and Workers. Unless otherwise specified, the Cost of Living Index is based on list prices, negotiated prices and free market prices.

### Free Market Price Index
Free Market Price Indexes refer to that of urban areas, unless otherwise specified.

Free Market Price Index of Urban Areas covers consumer goods only, and Free Market Price Index of Rural Areas covers consumer goods and agricultural producer goods.

# CHINA STATISTICAL ABSTRACT 1988

*Section I*

# GENERAL SURVEY

## 1.1 *ADMINISTRATIVE DIVISIONS*

(year-end 1987)

| **Provinces, Autonomous Regions and Municipalities Directly under the Central Government** | **Number of Prefectures** | **Number of Cities** | | | **Number of Counties** | **Number of Districts Under the Administration of Cities** |
|---|---|---|---|---|---|---|
| | | Total | Prefecture Level | County Level | | |
| ***Total*** | **156** | ***378** | **170** | **208** | **1986** | **632** |
| *Beijing Municipality* | | | | | 8 | 10 |
| *Tianjin Muncipality* | | | | | 5 | 13 |
| *Hebei Province* | 9 | 18 | 9 | 9 | 131 | 36 |
| *Shanxi Province* | 6 | 10 | 5 | 5 | 96 | 17 |
| *Inner Mongolia Autonomous Region* | 8 | 16 | 4 | 12 | 72 | 16 |
| *Liaoning Province* | | 19 | 13 | 6 | 39 | 55 |
| *Jilin Province* | 2 | 14 | 6 | 8 | 33 | 18 |
| *Heilongjiang Province* | 4 | 19 | 10 | 9 | 60 | 63 |
| *Shanghai Municipality* | | | | | 10 | 12 |
| *Jiangsu Province* | | 21 | 11 | 10 | 54 | 42 |
| *Zhejiang Province* | 2 | 18 | 9 | 9 | 58 | 22 |
| *Anhui Province* | 7 | 16 | 9 | 7 | 65 | 35 |
| *Fujian Province* | 3 | 11 | 6 | 5 | 59 | 17 |
| *Jiangxi Province* | 5 | 12 | 6 | 6 | 77 | 17 |
| *Shangdong Province* | 5 | 25 | 10 | 15 | 86 | 33 |
| *Henan Province* | 5 | 18 | 12 | 6 | 111 | 40 |
| *Hubei Province* | 7 | 25 | 8 | 17 | 54 | 27 |
| *Hunan Province* | 7 | 24 | 6 | 18 | 80 | 24 |
| *Guangdong Province* | 4 | 18 | 11 | 7 | 92 | 27 |
| *Guangxi Zhuang Autonomous Region* | 8 | 11 | 5 | 6 | 77 | 21 |
| *Sichuan Province* | 9 | 19 | 11 | 8 | 174 | 31 |
| *Guizhou Province* | 7 | 8 | 2 | 6 | 75 | 6 |
| *Yunnan Province* | 15 | 11 | 2 | 9 | 114 | 4 |
| *Tibet Autonomous Region* | 6 | 2 | 1 | 1 | 76 | 1 |
| *Shaanxi Province* | 6 | 8 | 4 | 4 | 89 | 14 |
| *Gansu Province* | 9 | 13 | 5 | 8 | 67 | 10 |
| *Qinghai Province* | 7 | 2 | 1 | 1 | 37 | 4 |
| *Ningxia Hui Autonomous Region* | 2 | 4 | 2 | 2 | 16 | 6 |
| *Xinjiang Uygur Autonomous Region* | 13 | 16 | 2 | 14 | 71 | 11 |
| *Taiwan Province* | | | | | | |

**Note:**
*** excluding Beijing, Tianjin and Shanghai.**

## 1.2 *POPULATION AND NATURAL RESOURCES**

| Item | | 1987 |
|---|---|---|
| ***Population (year end)*** ** | 1,000,000 persons | 1080.73 |
| ***Social Labor Force (year end)*** | 1,000,000 persons | 527.83 |
| ***National Surface Area*** | 1,000,000 square kilometers | 9.60 |
| ***Average Annual Rainfall*** | millimeters | 630 |
| ***Volume of Annual Rainfall*** | 1,000,000,000 cubic meters | 6000 |
| ***Afforested Areas*** | 1,000,000 hectares | 115.25 |
| ***Percentage of Forest Areas*** | % | 12.0 |
| ***Timber Reserves*** | 1,000,000,000 cubic meters | 10.26 |
| ***Grassland Areas*** | 1,000,000 hectares | 319.08 |
| *Utilizable Areas* | 1,000,000 hectares | 224.34 |
| ***Annual Flow of Rivers*** | 1,000,000,000 cubic meters | 2614.4 |
| ***Freshwater Areas*** | 1,000,000 hectares | 166.4 |
| *Cultivable Areas* | 1,000,000 hectares | 5.03 |
| *Cultivated Areas* | 1,000,000 hectares | 3.05 |
| ***Hydropower Resources*** | 1,000,000,000 kilowatts | 0.676 |
| *Developable Resources* | 1,000,000,000 kilowatts | 0.397 |
| ***Marine Fishing Areas*** | 1,000,000 square nautical miles | 0.818 |
| ***Cultivable Areas in Marine Water*** | 1,000,000 hectares | 0.492 |
| *Cultivated Areas* | 1,000,000 hectares | 0.163 |
| ***Length of Mainland Coastline*** | kilometers | 18000 |
| ***Coal Resources*** *** | 1,000,000,000 tons | 845.86 |
| ***Iron Ore Resources*** *** | 1,000,000,000 tons | 49.25 |

**Note:**
*** The figures for grassland and water areas and hydropower resources are obtained from statistical checks of previous years, and therefore are subject to further census and survey.**

**** Including military personnel.**

***** Coal and iron resources are 1985 figures.**

## 1.3 *MAIN INDICATORS OF NATIONAL ECONOMY**

| Item | | 1978 | 1980 | 1985 | 1986 | 1987 |
|---|---|---|---|---|---|---|
| **Population** | 1,000,000 persons | 962.59 | 987.05 | 1050.44 | 1065.29 | 1080.73 |
| **Social Labor Force** ** | 1,000,000 persons | 398.56 | 418.96 | 498.73 | 512.82 | 527.83 |
| Staff and Workers | 1,000,000 persons | 94.99 | 104.44 | 123.58 | 128.09 | 132.14 |
| **Total Ouput Value** | 1,000,000,000 yuan | 684.6 | 853.1 | 1658.7 | 1906.6 | 2279.3 |
| Gross Output Value of Agriculture and Industry | 1,000,000,000 yuan | 563.3 | 706.6 | 1333.7 | 1520.7 | 1848.2 |
| Gross Output Value of Agriculture | 1,000,000,000 yuan | 139.7 | 192.3 | 362.0 | 401.3 | 467.6 |
| Gross Output Value of Industry | 1,000,000,000 yuan | 423.6 | 514.3 | 971.7 | 1119.4 | 1380.6 |
| *Gross Output Value of Light Industry**** | 1,000,000,000 yuan | 182.6 | 242.5 | 461.0 | 533.0 | 661.2 |
| *Gross Output Value of Heavy Industry**** | 1,000,000,000 yuan | 241.0 | 271.8 | 510.7 | 586.4 | 719.4 |
| **National Income** | 1,000,000,000 yuan | 301.0 | 368.8 | 703.1 | 789.4 | 915.3 |
| **Gross Domestic Product** | 1,000,000,000 yuan | | | 833.0 | 946.4 | 1092.0 |
| **Government Revenue** | 1,000,000,000 yuan | 112.11 | 108.52 | 186.64 | 226.03 | 234.66 |
| Government Expenditure | 1,000,000,000 yuan | 111.10 | 121.27 | 184.48 | 233.08 | 242.69 |
| **Investments in Fixed Assets by State Owned Enterprises** | 1,000,000,000 yuan | 66.87 | 74.59 | 168.05 | 197.85 | 226.2 |
| **Volume of Freight Traffic** | 1,000,000,000 ton-km | 982.9 | 1202.6 | 1812.6 | 2014.8 | 2192.7 |
| **Cargo Handled at Main Seaports** | 1,000,000 ton | 198.34 | 217.31 | 311.54 | 344.09 | 365.70 |
| **Postal and Telecommunication Services** | 1,000,000,000 yuan | 1.17 | 1.33 | 2.96 | 3.29 | 3.88 |
| **Total Value of Retail Sales** | 1,000,000,000 yuan | 155.86 | 214.00 | 430.50 | 495.00 | 582.00 |
| **Total Value of Imports and Exports** | 1,000,000,000 U. S. dollars | 20.64 | 38.14 | 69.6 | 78.83 | 82.73 |
| Exports | 1,000,000,000 U. S. dollars | 9.75 | 18.12 | 27.35 | 30.94 | 39.49 |
| Imports | 1,000,000,000 U. S. dollars | 10.89 | 20.02 | 42.25 | 42.91 | 43.24 |
| **Output of Major Industrial and Farm Products** | | | | | | |
| Cloth | 1,000,000,000 m | 11.03 | 13.47 | 14.67 | 16.47 | 16.71 |
| Machine-made Paper and Paperboards | 1,000,000 ton | 4.39 | 5.35 | 9.11 | 9.99 | 10.08 |
| Coal | 1,000,000 ton | 0.618 | 0.62 | 0.872 | 0.894 | 0.928 |
| Electricity | 1,000,000,000 Kwh | 256.6 | 300.6 | 410.7 | 449.5 | 497.3 |
| Crude Oil | 1,000,000 ton | 104.05 | 105.95 | 124.90 | 130.69 | 134.04 |
| Steel | 1,000,000 ton | 31.78 | 37.12 | 46.79 | 52.2 | 56.02 |
| Rolled Steel Final Products | 1,000,000 ton | 22.08 | 27.16 | 36.93 | 40.58 | 43.91 |
| Cement | 1,000,000 ton | 65.24 | 79.86 | 145.95 | 166.06 | 181.28 |
| Grain | 1,000,000 ton | 304.77 | 320.56 | 379.11 | 391.51 | 404.73 |
| Cotton | 1,000,000 ton | 2.167 | 2.707 | 4.147 | 3.54 | 4.245 |
| Oil Bearing Crops | 1,000,000 ton | 5.218 | 7.691 | 15.784 | 14.738 | 15.278 |
| Raw Sugar | 1,000,000 ton | 23.818 | 29.112 | 60.468 | 58.525 | 55.504 |
| Pork, Beef and Mutton | 1,000,000 ton | 8.563 | 12.054 | 17.607 | 19.171 | 19.86 |
| Aquatic Products | 1,000,000 ton | 4.66 | 4.50 | 7.05 | 8.24 | 9.53 |

**Note:**

*** Figures in value terms in this table except postal and telecommunication services are at current prices. Postal and telecommunication services figures before 1981 are at 1970 constant prices, since 1981 they are at 1980 constant prices.**

**** Social labor force includes all employees and self-employ in urban and rural areas, i. e.: 1) staff and workers of state owned units (including government agencies) in all areas; 2) staff and workers of collective owned units in urban areas; 3) staff and workers of other ownership units in urban areas; 4) individual laborers in urban areas; and 5) collective and individual laborers in rural areas.**

***** Light industry produces consumer goods and hand tools. Heavy industry produces producer goods.**

## 1.4 *INDEX AND AVERAGE ANNUAL PERCENTAGE INCREASE RATE OF MAIN INDICATORS OF NATIONAL ECONOMY*

| Item | 1987 as Percentage of the Following Years | | | Average Annual Percentage Increase | |
|---|---|---|---|---|---|
| | 1978 | 1980 | 1986 | 1979-1987 | 1981-1987 |
| **Population** | 112.3 | 109.5 | 101.4 | 1.3 | 1.3 |
| **Social Labor Force** | 132.4 | 126.0 | 102.9 | 3.2 | 3.4 |
| Staff and Workers | 139.1 | 126.5 | 103.2 | 3.7 | 3.4 |
| **Total Ouput Value*** | 248.5 | 211.3 | 113.1 | 10.6 | 11.3 |
| Gross Output Value of Agriculture and Industry | 247.2 | 213.0 | 115.1 | 10.6 | 11.4 |
| Gross Output Value of Agriculture | 177.1 | 162.3 | 105.9 | 6.6 | 7.2 |
| Gross Output Value of Industry | 275.9 | 232.0 | 117.7 | 11.9 | 12.8 |
| *Gross Output Value of Light Industry* | 326.1 | 249.3 | 118.0 | 14.0 | 13.9 |
| *Gross Output Value of Heavy Industry* | 238.5 | 216.7 | 117.5 | 10.1 | 11.7 |
| **National Income*** | 214.9 | 188.7 | 109.3 | 8.9 | 9.5 |
| **Gross Domestic Product*** | | | 109.4 | | 9.8 |
| **Goverment Revenue** | 209.3 | 216.2 | 103.8 | 8.6 | 11.6 |
| Government Expenditure | 218.4 | 200.1 | 104.1 | 9.0 | 10.4 |
| **Investments in Fixed Assets by State Owned Enterprises** | 338.3 | 303.3 | 114.3 | 14.5 | 17.2 |
| **Volume of Freight Traffic** | 223.1 | 182.3 | 108.8 | 9.3 | 9.0 |
| **Cargo Handled at Main Seaports** | 184.4 | 168.3 | 106.3 | 7.0 | 7.7 |
| **Postal and Telecommunication Services*** | 331.6 | 291.7 | 117.9 | 14.2 | 16.5 |
| **Total Value of Retail Sales** | 373.4 | 272.0 | 117.6 | 15.8 | 15.4 |
| **Total Value of Imports and Exports** | 400.8 | 216.9 | 112.0 | 16.7 | 11.7 |
| Exports | 405.0 | 217.9 | 127.6 | 16.8 | 11.8 |
| Imports | 397.1 | 216.0 | 100.7 | 16.6 | 11.6 |
| **Output of Major Industrial and Farm Products** | | | | | |
| Cloth | 151.5 | 124.1 | 101.5 | 4.7 | 3.1 |
| Machine-made Paper and Paper-boards | 229.6 | 188.4 | 100.9 | 9.7 | 9.5 |
| Coal | 150.2 | 149.7 | 103.8 | 4.6 | 5.9 |
| Electricity | 193.8 | 165.4 | 110.6 | 7.6 | 7.5 |
| Crude Oil | 128.8 | 126.5 | 102.6 | 2.9 | 3.4 |
| Steel | 176.3 | 150.9 | 107.3 | 6.5 | 6.1 |
| Rolled Steel Final Products | 198.9 | 161.7 | 108.2 | 7.9 | 7.1 |
| Cement | 277.9 | 227.0 | 109.2 | 12.0 | 12.4 |
| Grain | 132.8 | 126.3 | 103.4 | 3.2 | 3.4 |
| Cotton | 195.9 | 156.8 | 119.9 | 7.8 | 6.6 |
| Oil Bearing Crops | 292.8 | 198.6 | 103.7 | 12.7 | 10.3 |
| Crude Sugar | 233.0 | 190.7 | 94.8 | 9.9 | 9.7 |
| Pork, Beef and Mutton | 231.9 | 164.8 | 103.6 | 9.8 | 7.4 |
| Aquatic Products | 204.5 | 211.8 | 155.7 | 8.3 | 11.3 |

**Note:**
*** The indexes and average annual percentage increase for these items are based on comparable prices.**

## 1.5 *COMPOSITION OF SELECTED AGGREGATE STATISTICS**

| | 1978 | 1980 | 1986 | 1987 |
|---|---|---|---|---|
| ***Percentage of Gross Ouput Value of Agriculture and Industry*** | | | | |
| *Agriculture* | 24.8 | 27.2 | 26.4 | 25.3 |
| *Light Industry* | 32.4 | 34.3 | 35.0 | 35.8 |
| *Heavy Industry* | 42.8 | 38.5 | 38.6 | 38.9 |
| ***Percentage of Gross Industrial Output Value*** | | | | |
| *Light Industry* | 43.1 | 47.2 | 47.6 | 47.9 |
| *Heavy Industry* | 46.9 | 52.8 | 52.4 | 52.1 |
| ***Percentage of Gross Agricultural Output Value*** | | | | |
| *Farming (crop cultivation)* | 76.7 | 71.7 | 62.3 | 60.7 |
| *Forestry* | 3.4 | 4.2 | 5.0 | 4.7 |
| *Animal Husbandry* | 15.0 | 18.4 | 21.8 | 22.8 |
| *Sideline Occupations* | 3.3 | 4.0 | 6.9 | 7.0 |
| *Fishery* | 1.6 | 1.7 | 4.1 | 4.8 |
| ***Percentage of National Income Available*** | | | | |
| *Accumulation* | 36.5 | 31.5 | 34.8 | 34.4 |
| *Consumption* | 63.5 | 68.5 | 65.2 | 65.6 |
| ***Percentage of Projects in Capital Construction**** | | | | |
| *Investment in Productive Projects* | 79.1 | 64.3 | 60.6 | 65.9 |
| *Investment in Non-Productive Projects* | 20.9 | 35.7 | 39.4 | 34.1 |
| *Housing Projects* | 7.8 | 20.0 | 16.1 | 13.4 |
| ***Percentage of Investment in Capital Construction**** | | | | |
| *Investment in Agriculture* | 10.6 | 9.3 | 3.0 | 4.8 |
| *Investment in Light Industry* | 5.8 | 9.1 | 7.0 | 5.1 |
| *Investment in Heavy Industry* | 48.7 | 40.2 | 38.2 | 46.4 |
| ***Percentage of Energy Industry and Transportation in Capital Construction*** | | | | |
| *Investment in Energy Industry* | 22.9 | 20.7 | 22.5 | 23.3 |
| *Investment in Transportation, Postal and Telecommunication Services* | 13.6 | 11.2 | 15.4 | 15.4 |
| ***Government Revenue as Percentage of National Income***** | 28.3 | 27.7 | 24.5 | 24.5 |
| ***Appropriations for Capital Construction as Percentage of Government Expenditure*** | 40.7 | 34.6 | 28.8 | 25.2 |
| ***Expenditure on Culture, Education, Public Health and Science as Percentage of Government Expenditure*** | 10.1 | 12.9 | 16.3 | 16.7 |

**Note:**
*** Figures in this table are based on current prices.**

**** Productive projects are projects for material production or projects directly serving material production. Non-productive projects refer to residential buildings, scientific research institutions, cultural, educational and healthcare facilities, public services, and other administrative agencies that serve the needs of people.**

***** Excluding foreign loans as government revenue.**

## 1.6 *TOTAL OUTPUT VALUE OF SOCIETY AND NATIONAL INCOME*

| Year | | National Income* | | Index (1978 = 100) | |
|---|---|---|---|---|---|
| | Total Ouput Value (1,000,000,000 yuan) | Total National Income (1,000,000,000 yuan) | Per Capita National Income (yuan) | Total Ouput Value | National Income |
| *1978* | 684.6 | 301.0 | 315 | 100.0 | 100.0 |
| *1979* | 764.2 | 335.0 | 346 | 108.5 | 107.0 |
| *1980* | 853.1 | 368.8 | 376 | 117.6 | 113.9 |
| *1981* | 907.1 | 394.0 | 396 | 123.0 | 119.4 |
| *1982* | 996.3 | 426.1 | 432 | 134.7 | 129.3 |
| *1983* | 1112.5 | 473.0 | 463 | 148.5 | 142.0 |
| *1984* | 1316.7 | 565.0 | 547 | 170.3 | 161.2 |
| *1985* | 1658.7 | 703.1 | 673 | 199.7 | 182.3 |
| *1986* | 1906.6 | 789.4 | 746 | 219.8 | 196.6 |
| *1987* | 2279.3 | 915.3 | 853 | 248.5 | 214.9 |

**Note:**
*** The national income figures in this table are based on current prices and indexes are at comparable prices.**

## 1.7 *NATIONAL INCOME, ACCUMULATION, GOVERNMENT REVENUE AND EXPENDITURE, AND ALLOCATIONS FOR CAPITAL CONSTRUCTION*

| Year | Accumulation as Percentage of National Income Available | Government Revenue as Percentage of National Income | | Allocations for Capital Construction as Percentage of Governement Expenditure |
|---|---|---|---|---|
| | | Including Foreign Borrowing | Excluding Foreign Borrowing | |
| *1978-1986* | 32.6 | 28.5 | | 32.2 |
| *1981-1986* | 32.2 | 27.1 | 26.3 | 29.8 |
| *1978* | 36.5 | 37.2 | | 40.7 |
| *1979* | 34.6 | 32.9 | 31.9 | 40.4 |
| *1980* | 31.5 | 29.4 | 28.3 | 34.6 |
| *1981* | 28.3 | 27.7 | 25.8 | 29.7 |
| *1982* | 28.8 | 26.4 | 25.4 | 26.8 |
| *1983* | 29.7 | 26.4 | 25.6 | 29.6 |
| *1984* | 31.5 | 26.6 | 26.0 | 31.6 |
| *1985* | 35.2 | 26.5 | 26.1 | 31.6 |
| *1986* | 34.8 | 28.6 | 27.7 | 28.8 |
| *1987* | 34.4 | 25.6 | 24.5 | 25.2 |

## 1.8 GROSS OUTPUT VALUE OF AGRICULTURE AND INDUSTRY *

(1,000,000,000 yuan )

| Year | Gross Output Value of Agriculture and Industry | Gross Output Value of Agriculture | Gross Output Value of Industry ** | | |
|---|---|---|---|---|---|
| | | | Total | Light Industry | Heavy Industry |
| 1978 | 563.3 | 139.7 | 423.6 | 182.6 | 241.0 |
| 1979 | 638.0 | 169.8 | 468.2 | 204.5 | 203.7 |
| 1980 | 706.6 | 192.3 | 514.3 | 242.5 | 271.8 |
| 1981 | 757.5 | 218.1 | 539.4 | 277.8 | 261.6 |
| 1982 | 829.4 | 248.3 | 531.1 | 291.9 | 289.2 |
| 1983 | 920.6 | 275.0 | 645.6 | 323.2 | 332.4 |
| 1984 | 1083.1 | 321.4 | 761.7 | 360.8 | 400.9 |
| 1985 | 1333.7 | 362.0 | 971.7 | 461.0 | 510.7 |
| 1986 | 1520.7 | 401.3 | 1119.4 | 533.0 | 586.4 |
| 1987 | 1848.2 | 467.6 | 1380.6 | 661.2 | 719.4 |

Note:
* Figures in this table are at current prices.

** Including village-run industry.

## 1.9 COMPOSITION OF GROSS OUTPUT VALUE OF AGRICULTURE AND INDUSTRY*

| Year | Percentage of the Gross Output Value of Agriculture and Industry | | | Percentage of the Gross Output Value of Industry | |
|---|---|---|---|---|---|
| | Agriculture | Light Industry | Heavy Industry | Light Industry | Heavy Industry |
| 1978 | 24.8 | 32.4 | 42.8 | 43.1 | 56.9 |
| 1979 | 26.6 | 32.1 | 41.3 | 43.7 | 56.3 |
| 1980 | 27.2 | 34.3 | 38.5 | 47.2 | 52.8 |
| 1981 | 28.8 | 36.7 | 35.4 | 51.5 | 48.5 |
| 1982 | 29.9 | 35.2 | 34.9 | 50.2 | 49.8 |
| 1983 | 29.9 | 34.0 | 36.1 | 48.5 | 51.5 |
| 1984 | 29.7 | 33.3 | 37.0 | 47.4 | 52.6 |
| 1985 | 27.1 | 34.6 | 38.3 | 47.4 | 52.6 |
| 1986 | 26.4 | 35.0 | 38.6 | 47.6 | 52.4 |
| 1987 | 25.3 | 35.8 | 38.9 | 47.9 | 52.1 |

Note:
* Figures in this table are based on current prices.

## 1.10 *INDEXES OF GROSS OUTPUT VALUE OF AGRICULTURE AND INDUSTRY**

(1978 = 100)

| Year | Total Gross Output Value | Gross Output Value of Agriculture | Gross Output Value of Industry ** | | |
|---|---|---|---|---|---|
| | | | Total | Light Industry | Heavy Industry |
| 1978 | 100.00 | 100.00 | 100.00 | 100.00 | 100.00 |
| 1979 | 108.50 | 107.60 | 108.80 | 109.90 | 108.00 |
| 1980 | 116.10 | 109.10 | 118.90 | 130.70 | 110.10 |
| 1981 | 121.70 | 116.20 | 124.00 | 149.40 | 105.00 |
| 1982 | 132.40 | 129.30 | 133.70 | 158.10 | 115.50 |
| 1983 | 145.90 | 131.30 | 148.60 | 172.80 | 130.60 |
| 1984 | 168.10 | 156.40 | 172.90 | 200.70 | 152.00 |
| 1985 | 195.80 | 161.80 | 209.80 | 246.20 | 182.70 |
| 1986 | 214.80 | 167.30 | 234.30 | 276.30 | 203.00 |
| 1987 | 247.20 | 177.10 | 275.90 | 326.10 | 238.50 |

### *INDEXES OF GROSS OUTPUT VALUE OF AGRICULTURE AND INDUSTRY **

(preceding year = 100)

| Year | Gross Output Value of Agriculture and Industry | Gross Output Value of Agriculture | Gross Output Value of Industry ** | | |
|---|---|---|---|---|---|
| | | | Total | Light Industry | Heavy Industry |
| 1978 | 112.3 | 108.1 | 113.6 | 110.9 | 115.6 |
| 1979 | 108.5 | 107.6 | 108.8 | 109.9 | 108.0 |
| 1980 | 107.0 | 101.4 | 109.3 | 119.0 | 101.9 |
| 1981 | 104.9 | 106.5 | 104.3 | 114.3 | 95.4 |
| 1982 | 108.8 | 111.3 | 107.8 | 105.8 | 109.9 |
| 1983 | 110.2 | 107.7 | 111.2 | 109.3 | 113.1 |
| 1984 | 115.2 | 112.3 | 116.3 | 116.1 | 116.5 |
| 1985 | 116.5 | 103.5 | 121.4 | 122.6 | 120.2 |
| 1986 | 109.7 | 103.4 | 111.7 | 112.2 | 111.1 |
| 1987 | 115.1 | 105.9 | 117.7 | 118.0 | 117.5 |

**Note:**
*** Figures in this table are based on comparable prices.**

**** Including village-run industry.**

## 1.11 *GROSS OUPUT VALUE OF AGRICULTURE AND INDUSTRY BY REGION 1987 **

(1,000,000,000 yuan)

| Region | Total Gross Output Value | Gross Output Value of Agriculture | Gross Output Value of Industry | Percentage of Gross Output Value of Industry | |
|---|---|---|---|---|---|
| | | | | Light Industry | Heavy Industry |
| ***National Total*** | **1848.21** | **467.57** | **1380.64** | **25.3** | **74.7** |
| *Beijing* | 46.76 | 3.44 | 43.32 | 7.4 | 92.6 |
| *Tianjin* | 44.01 | 3.29 | 40.72 | 7.5 | 92.5 |
| *Hebei* | 84.16 | 20.07 | 64.09 | 23.8 | 76.2 |
| *Shanxi* | 36.57 | 6.14 | 30.43 | 16.8 | 83.2 |
| *Inner Mongolia* | 23.88 | 8.77 | 15.11 | 36.7 | 63.3 |
| *Liaoning* | 121.33 | 16.92 | 104.41 | 14.0 | 86.0 |
| *Jilin* | 48.02 | 12.08 | 35.94 | 25.2 | 74.8 |
| *Heilongjiang* | 70.10 | 13.70 | 56.40 | 19.5 | 80.5 |
| *Shanghai* | 110.61 | 3.88 | 106.73 | 3.5 | 96.5 |
| *Jiangsu* | 196.96 | 38.03 | 158.93 | 19.3 | 80.7 |
| *Zhejiang* | 107.88 | 22.81 | 85.07 | 21.1 | 78.9 |
| *Anhui* | 64.94 | 25.53 | 39.41 | 39.3 | 60.7 |
| *Fujian* | 39.69 | 13.30 | 26.39 | 33.5 | 66.5 |
| *Jiangxi* | 40.29 | 14.44 | 25.85 | 35.8 | 64.2 |
| *Shangdong* | 144.61 | 41.32 | 103.29 | 28.6 | 71.4 |
| *Henan* | 92.03 | 32.36 | 59.67 | 35.2 | 64.8 |
| *Hubei* | 90.88 | 24.97 | 65.91 | 27.5 | 72.5 |
| *Hunan* | 71.01 | 25.34 | 45.67 | 35.7 | 64.3 |
| *Guangdong* | 128.08 | 38.99 | 89.09 | 30.4 | 69.6 |
| *Guangxi* | 34.54 | 13.79 | 20.75 | 39.9 | 60.1 |
| *Sichuan* | 111.39 | 38.89 | 72.51 | 34.9 | 65.1 |
| *Guizhou* | 21.90 | 9.23 | 12.67 | 42.1 | 57.9 |
| *Yunnan* | 29.30 | 11.13 | 18.17 | 38.0 | 62.0 |
| *Tibet* | 1.24 | 1.03 | 0.21 | 83.1 | 16.9 |
| *Shaanxi* | 36.20 | 10.34 | 25.86 | 28.6 | 71.4 |
| *Gansu* | 22.54 | 6.55 | 15.99 | 29.1 | 70.9 |
| *Qinghai* | 4.71 | 1.59 | 3.12 | 33.8 | 66.2 |
| *Ningxia* | 4.82 | 1.48 | 3.34 | 30.7 | 69.3 |
| *Xinjiang* | 19.76 | 8.16 | 11.60 | 41.3 | 58.7 |

**Note:**
*** Figures in this table are at current prices.**

# 1.12 *SELECTED INDICATORS OF AVERAGE DAILY SOCIAL AND ECONOMIC ACTIVITIES **

| Item | | 1980 | 1987 |
|---|---|---|---|
| **Daily Production** | | | |
| Total Output Value of Society | 1,000,000,000 yuan | 2.34 | 6.24 |
| Gross Output Value of Agriculture and Industry ** | 1,000,000,000 yuan | 1.94 | 5.06 |
| *Gross Agricultural Output Value* | 1,000,000,000 yuan | 0.60 | 1.28 |
| *Gross Industrial Output Value ** * | 1,000,000,000 yuan | 1.34 | 3.78 |
| National Income | 1,000,000,000 yuan | 1.01 | 2.51 |
| Government Revenue | 1,000,000,000 yuan | 0.30 | 0.64 |
| Cloth | 1,000,000 meters | 36.90 | 45.78 |
| Machine-made Paper and Paperboards | 1,000,000 ton | 0.015 | 0.028 |
| Coal | 1,000,000 ton | 1.699 | 2.542 |
| Electricity | 1,000,000,000 Kwh | 0.82 | 1.36 |
| Crude Oil | 1,000,000 ton | 0.290 | 0.367 |
| Steel | 1,000,000 ton | 0.102 | 0.153 |
| Rolled Steel Final Products | 1,000,000 ton | 0.074 | 0.120 |
| Cement | 1,000,000 ton | 0.219 | 0.497 |
| **Daily National Consumption** | | | |
| Total Consumption of Urban and Rural Residents | 1,000,000,000 yuan | 0.61 | 1.48 |
| Per Capita Consumption | yuan | 0.6 | 1.4 |
| Grain | 1,000,000 ton | 0.575 | 0.739 |
| Pork | 1,000,000 ton | 0.030 | 0.043 |
| Edible Vegetable Oil | 1,000,000 ton | 0.006 | 0.016 |
| Sugar | 1,000,000 ton | 0.010 | 0.020 |
| Cloth for Daily Use | 1,000,000 meters | 26.94 | 33.23 |
| **Daily Retail Sales** | | | |
| Bicycles | 1,000,000 | 0.032 | 0.098 |
| Wrist Watches | 1,000,000 | 0.069 | 0.149 |
| Television Sets | 1,000,000 | 0.010 | 0.064 |
| Refrigerators | unit | 151 | 11241 |
| Washing Machines | 1,000,000 | 0.0006 | 0.039 |
| **Other Daily Economic Activities** | | | |
| New Residential Buildings in Urban Areas | 1,000,000 square meters | 0.252 | 0.493 |
| New Residential Buildings Completed in Rural Areas | 1,000,000 square meters | 1.370 | 2.219 |
| Books Published | 1,000,000 copies | 12.58 | 17.12 |
| Magazines Issued | 1,000,000 copies | 3.07 | 7.29 |
| Newspaper Circulated | 1,000,000 copies | 38.47 | 57.32 |
| Posted Letters | 1,000,000 | 9.08 | 15.01 |
| **Daily Population Changes and Marital Events** | | | |
| Births | 1,000,000 persons | 0.049 | 0.062 |
| Deaths | 1,000,000 persons | 0.017 | 0.019 |
| Marriages | 1,000,000 couples | 0.020 | 0.025 |
| Divorces | couples | 494 | 1592 |

**Note:**
*** Figures in value terms in this table are at current prices.**

**** Industries run by villages and below are included in industry.**

## 1.13 *ORIGINAL VALUE OF FIXED ASSETS OF STATE OWNED ENTERPRISES* *

(1,000,000,000 yuan) (year end figures)

| Sector | 1978 | 1980 | 1984 | 1985 | 1986 |
|---|---|---|---|---|---|
| ***Total*** | **448.82** | **531.11** | **737.05** | **800.49** | **904.18** |
| *Industry* | 300.22 | 346.52 | 480.64 | 518.22 | 587.17 |
| *Farming, Animal Husbandry, Fishery and Forestry* | 12.68 | 16.75 | 20.35 | 33.92 | 37.57 |
| *Construction* ** | 7.26 | 13.64 | 20.31 | 22.43 | 26.91 |
| *Transportation, Postal and Telecommunication Services* | 82.41 | 94.33 | 120.12 | 132.78 | 146.40 |
| *Commerce, Grain Supply and Foreign Trade* | 31.58 | 40.96 | 65.32 | 56.67 | 64.04 |
| *Supply and Marketing of Producer Goods* | 4.02 | 5.76 | 9.02 | 12.20 | 14.77 |
| *Urban Public Utilities* | 5.23 | 6.52 | 10.76 | 12.39 | 14.21 |
| *Culture, Education, Science and Public Health* | 2.32 | 2.86 | 4.79 | 5.35 | 6.15 |

**Note:**
*** Figures in this table are classified by administrative departments of various sectors in the central government of the People's Republic of China.**

**** Figures in the construction sector exclude private managed construction units.**

## 1.14 *TARGET CIRCULATION FUNDS OF STATE OWNED ENTERPRISES* *

(1,000,000,000 yuan) (year end figures)

| Sector | 1978 | 1980 | 1984 | 1985 | 1986 |
|---|---|---|---|---|---|
| ***Total*** | **285.34** | **322.28** | **391.51** | **396.98** | **456.65** |
| *Industry* | 95.94 | 99.70 | 128.83 | 150.58 | 180.57 |
| *Farming, Animal Husbandry, Fishery and Forestry* | 12.24 | 12.39 | 9.27 | 13.04 | 15.28 |
| *Construction* ** | 3.86 | 5.76 | 10.06 | 13.03 | 16.96 |
| *Transportation, Posts and Telecommunications* | 7.22 | 6.91 | 7.01 | 8.70 | 10.16 |
| *Commerce, Grain Supply and Foreign Trade* | 130.00 | 150.96 | 184.07 | 141.60 | 161.91 |
| *Supply and Marketing of Producer Goods* | 32.60 | 41.42 | 46.82 | 62.82 | 64.26 |
| *Urban Public Utilities* | 0.48 | 0.49 | 0.62 | 0.82 | 0.93 |
| *Culture, Education, Science and Public Health* | 1.39 | 1.79 | 2.67 | 3.84 | 3.92 |

**Note:**
*** Figures in this table are classified by administrative departments of various sectors in the central government of the People's Republic of China.**

**** Figures in the construction sector exclude privately managed construction units.**

## 1.15 *MAIN ECONOMIC STATISTICS OF FOURTEEN COASTAL CITIES 1987 **

| Item | | Dalian | Qinhuangdao | Tianjin | Yantai | Qingdao |
|---|---|---|---|---|---|---|
| ***Gross Output Value of Industry*** | 1,000,000,000 yuan | 12.884 | 1.561 | 32.586 | 6.679 | 12.881 |
| *Joint Ventures, Cooperative Operation and Foreign Capital Enterprise* | 1,000,000,000 yuan | 0.029 | 0.053 | 0.430 | | 0.017 |
| ***Cargo Handled at Seaports*** | 1,000,000,000 ton | 46.10 | 53.79 | 17.21 | 10.83 | 30.70 |
| *Imports* | 1,000,000,000 ton | 10.15 | 2.97 | 10.55 | 5.46 | 7.01 |
| *Exports* | 1,000,000,000 ton | 35.95 | 50.82 | 6.67 | 5.37 | 23.69 |
| ***Investment in Fixed Assets of State Owned Units*** | 1,000,000,000 yuan | 3.547 | 0.912 | 6.138 | 1.226 | 1.510 |
| *Capital Construction* | 1,000,000,000 yuan | 2.418 | 0.675 | 1.794 | 0.864 | 0.771 |
| ***Total Value of Retail Sales*** | 1,000,000,000 yuan | 5.298 | 1.213 | 10.117 | 3.949 | 4.894 |
| ***Total Value of Purchases by Department of Foreign Trade*** | 1,000,000,000 yuan | 2.559 | 0.117 | 3.433 | 1.127 | 1.748 |
| ***Total Value of Exports at Trading Ports*** | 1,000,000 U. S. dollars | 4810.70 | 1506.67 | 3284.94 | 86.99 | 2673.10 |
| ***Number of Agreements for Utilization of Foreign Capital*** | | 61 | 6 | 249 | 45 | 24 |
| *Number of Agreements for Direct Investment* | | 31 | 6 | 35 | 15 | 14 |
| ***Total Value of Agreements for Utilization of Foreign Capital*** | 1,000,000 U. S. dollars | 188.33 | 21.59 | 404.95 | 30.9 | 65.91 |
| *Total Value for Direct Investment* | 1,000,000 U. S. dollars | 87.22 | 21.59 | 14.37 | 7.81 | 18.79 |
| ***Foreign Funds Already Invested*** | 1,000,000 U. S. dollars | 270.11 | 2.19 | 336.64 | 9.0 | 41.79 |
| *Direct Investment* | 1,000,000 U. S. dollars | 49.90 | 2.19 | 87.45 | 4.61 | 12.14 |
| ***Foreign Exchange Revenue (Bank of China)*** | 1,000,000 U. S. dollars | 278.86 | 620.91 | 1323.32 | 74.49 | 2456.59 |
| *Export Revenue* | 1,000,000 U. S. dollars | 206.97 | 596.63 | 1182.36 | 62.87 | 2414.20 |
| ***Local Government Revenue*** | 1,000,000,000 yuan | 0.222 | 0.0298 | 0.5294 | 0.0698 | 0.16 |
| ***Cost of Living Index of Staff and Workers*** | (previous year = 100) | 108.6 | 106.6 | 106.4 | 109.8 | 109.4 |

**Note:**
*** Statistics of fourteen coastal cities includes counties under the administration of cities.**

## MAIN ECONOMIC STATISTICS OF FOURTEEN COASTAL CITIES 1987 *

(continued)

| Item | | Lianyungang | Nantong | Shanghai | Ningbo | Wenzhou |
|---|---|---|---|---|---|---|
| **Gross Output Value of Industry** | 1,000,000,000 yuan | 2.579 | 11.721 | 92.308 | 12.195 | 4.332 |
| *Joint Ventures, Cooperative Operation and Foreign Capital Enterprise* | 1,000,000,000 yuan | 0.011 | 0.129 | 1.159 | 0.036 | 0.003 |
| **Cargo Handled at Seaports** | 1,000,000,000 ton | 89.4 | | 1288.4 | 194.0 | 45.4 |
| *Imports* | 1,000,000,000 ton | 33.8 | | 944.2 | 118.4 | 34.0 |
| *Exports* | 1,000,000,000 ton | 55.6 | | 344.3 | 75.6 | 11.4 |
| **Investment in Fixed Assets of State Owned Units** | 1,000,000,000 yuan | 1.006 | 0.853 | 14.969 | 0.952 | 0.322 |
| *Capital Construction* | 1,000,000,000 yuan | 0.789 | 0.760 | 9.232 | 0.641 | 0.239 |
| **Total Value of Retail Sales** | 1,000,000,000 yuan | 1.686 | 4.173 | 23.900 | 4.001 | 3.266 |
| **Total Value of Purchases by Department of Foreign Trade** | 1,000,000,000 yuan | 0.212 | 1.344 | 12.802 | 0.863 | 0.430 |
| **Total Value of Exports at Trading Ports** | 1,000,000 U. S. dollars | 2035.3 | 87.6 | 66003.8 | 2999.9 | |
| **Number of Agreements for Utilization of Foreign Capital** | | 8 | 36 | 187 | 22 | 4 |
| *Number of Agreements for Direct Investment* | | 4 | 12 | 76 | 13 | |
| **Total Value of Agreements for Utilization of Foreign Capital** | 1,000,000 U. S. dollars | 10.52 | 44.19 | 994.58 | 45.93 | 0.69 |
| *Total Value for Direct Investment* | 1,000,000 U. S. dollars | 2.65 | 16.32 | 338.32 | 42.42 | |
| **Foreign Funds Already Invested** | 1,000,000 U. S. dollars | 8.89 | 17.79 | 509.50 | 4.49 | 0.45 |
| *Direct Investment* | 1,000,000 U. S. dollars | 2.68 | 11.70 | 211.96 | 4.29 | |
| **Foreign Exchange Revenue (Bank of China)** | 1,000,000 U. S. dollars | 85.67 | 56.51 | 3996.46 | 15.34 | 5.42 |
| *Export Revenue* | 1,000,000 U. S. dollars | 77.38 | 50.78 | 3449.80 | 7.72 | 0.24 |
| **Local Government Revenue** | 1,000,000,000 yuan | 0.0279 | 0.0988 | 1.6509 | 0.1135 | 0.0613 |
| **Cost of Living Index of Staff and Workers** | (previous year = 100) | | 109.5 | 100.1 | 111.8 | 112.5 |

**Note:**
*** Statistics of fourteen coastal cities includes counties under the administration of cities.**

## MAIN ECONOMIC STATISTICS OF FOURTEEN COASTAL CITIES 1987 *
(continued)

| Item | | Fuzhou | Guangzhou | Zhanjiang | Beihai |
|---|---|---|---|---|---|
| ***Gross Output Value of Industry*** | 1,000,000,000 yuan | 5.183 | 20.049 | 2.443 | 0.541 |
| *Joint Ventures, Cooperative Operation and Foreign Capital Enterprise* | 1,000,000,000 yuan | 0.939 | 0.650 | 0.062 | 0.008 |
| ***Cargo Handled at Seaports*** | 1,000,000,000 ton | 50.6 | 209.1 | 170.0 | 7.4 |
| *Imports* | 1,000,000,000 ton | 28.7 | 148.3 | 115.9 | 5.3 |
| *Exports* | 1,000,000,000 ton | 21.9 | 60.8 | 54.1 | 2.1 |
| ***Investment in Fixed Assets of State Owned Units*** | 1,000,000,000 yuan | 1.405 | 4.931 | 0.770 | 0.178 |
| *Capital Construction* | 1,000,000,000 yuan | 1.140 | 3.000 | 0.421 | 0.082 |
| ***Total Value of Retail Sales*** | 1,000,000,000 yuan | 3.035 | 10.687 | 2.767 | 0.610 |
| ***Total Value of Purchases by Department of Foreign Trade*** | 1,000,000,000 yuan | 0.590 | 3.046 | 0.454 | 0.115 |
| ***Total Value of Exports at Trading Ports*** | 1,000,000 U. S. dollars | 3736.4 | 27127.0 | 10275.6 | 950.5 |
| ***Number of Agreements for Utilization of Foreign Capital*** | | 56 | 1545 | 48 | 13 |
| *Number of Agreements for Direct Investment* | | 56 | 137 | 33 | 13 |
| ***Total Value of Agreements for Utilization of Foreign Capital*** | 1,000,000 U. S. dollars | 30.52 | 291.65 | 44.09 | 18.51 |
| *Total Value for Direct Investment* | 1,000,000 U. S. dollars | 30.52 | 209.27 | 30.55 | 18.51 |
| ***Foreign Funds Already Invested*** | 1,000,000 U. S. dollars | 16.52 | 82.61 | 22.66 | 1.38 |
| *Direct Investment* | 1,000,000 U. S. dollars | 14.62 | 53.97 | 15.58 | 1.38 |
| ***Foreign Exchange Revenue (Bank of China)*** | 1,000,000 U. S. dollars | 19.53 | 516.48 | 70.38 | 75.75 |
| *Export Revenue* | 1,000,000 U. S. dollars | 13.67 | 439.09 | 51.68 | 74.45 |
| ***Local Government Revenue*** | 1,000,000,000 yuan | 0.0680 | 0.3445 | 0.0412 | 0.0092 |
| ***Cost of Living Index of Staff and Workers*** | (previous year = 100) | 110.6 | 113.7 | 118.2 | 112.7 |

**Note:**
*** Statistics of fourteen coastal cities includes counties under the administration of cities.**

*Section II*

# POPULATION AND LABOR FORCE

## 2.1 *NATIONAL POPULATION* *
(1,000,000 persons)

| Year | Total Population | Sex | | Residence | |
|---|---|---|---|---|---|
| | | Male | Female | **Urban | ***Rural |
| *1978* | 962.59 | 495.67 | 466.92 | 172.45 | 790.14 |
| *1979* | 975.42 | 501.92 | 473.50 | 184.95 | 790.47 |
| *1980* | 987.05 | 507.85 | 479.20 | 191.40 | 795.65 |
| *1981* | 1000.72 | 515.19 | 485.53 | 201.71 | 799.01 |
| *1982* | 1015.90 | 523.19 | 492.71 | 211.31 | 804.59 |
| *1983* | 1027.64 | 530.29 | 497.38 | 241.50 | 786.14 |
| *1984* | 1038.76 | 536.00 | 502.76 | 331.36 | 707.40 |
| *1985* **** | 1050.44 | 543.08 | 507.36 | 384.46 | 665.98 |
| *1986* | 1065.29 | 550.75 | 514.54 | 441.03 | 624.26 |
| *1987* | 1080.73 | 556.58 | 524.15 | 503.62 | 577.11 |

**Note:**
*** The figures in this table include military personnel.**

**** The urban figure refers to the population living in areas under the administration of cities and towns.**

***** The rural figure refers to the population of counties excluding those living in towns.**

****** After 1984, the increase of urban population is attributed mainly to the many newly established towns resulting from readjustment in the criteria for determining towns.**

## 2.2 *BIRTH RATE, DEATH RATE AND NATURAL GROWTH RATE*
(per 1,000)

| Year | Birth Rate | Death Rate | Natural Growth Rate |
|---|---|---|---|
| *1978* | 18.3 | 6.3 | 12.0 |
| *1979* | 17.8 | 6.2 | 11.6 |
| *1981** | 20.9 | 6.4 | 14.5 |
| *1982* | 21.1 | 6.6 | 14.5 |
| *1983* | 18.6 | 7.1 | 11.5 |
| *1984* | 17.5 | 6.7 | 10.8 |
| *1985* | 17.8 | 6.6 | 11.2 |
| *1986* | 20.8 | 6.7 | 14.1 |
| *1987* | 21.0 | 6.6 | 14.4 |

**Note:**
*** The data source of 1981 figure is from the 1982 population census. Other figures are from sample survey.**

## 2.3 *SOCIAL LABOR FORCE**

(1,000,000 persons) (year-end figure)

| Year | Social Labor Force | | | | Index (1978=100) |
|---|---|---|---|---|---|
| | Total | Staff and Workers | Individual Laborers in Cities and Towns | Collective/ Individual Laborers in Rural Areas | |
| *1978* | 398.56 | 94.99 | 0.15 | 303.42 | 1.00 |
| *1979* | 405.81 | 99.67 | 0.32 | 305.82 | 1.02 |
| *1980* | 418.96 | 104.44 | 0.81 | 313.71 | 1.05 |
| *1981* | 432.80 | 109.40 | 1.13 | 322.27 | 1.09 |
| *1982* | 447.06 | 112.81 | 1.47 | 332.78 | 1.12 |
| *1983* | 460.04 | 115.15 | 2.31 | 342.58 | 1.15 |
| *1984* | 475.97 | 118.90 | 3.39 | 353.68 | 1.19 |
| *1985* | 498.73 | 123.58 | 4.50 | 370.65 | 1.25 |
| *1986* | 512.83 | 128.09 | 4.83 | 379.90 | 1.29 |
| *1987* | 527.83 | 132.14 | 5.69 | 390.00 | 1.32 |

**Note:**
*** Social labor force includes all employed and self-employed in urban and rural areas, i. e., 1) staff and workers of state owned units (including government agencies) in all areas; 2) staff and workers of collective owned units in urban areas; 3) staff and workers of other ownership units in urban areas; 4) individual labors in urban areas; and 5) collective and individual labors in rural areas.**

## 2.4 *STAFF AND WORKERS**

(1,000,000 persons) (year-end figure)

| Year | Total | | | |
|---|---|---|---|---|
| | | State Owned Units | Collective Owned Units in Cities and Towns | Others |
| *1978* | 94.99 | 74.51 | 20.48 | |
| *1979* | 99.67 | 76.93 | 22.74 | |
| *1980* | 104.44 | 80.19 | 24.25 | |
| *1981* | 109.40 | 83.72 | 25.68 | |
| *1982* | 112.81 | 86.30 | 26.51 | |
| *1983* | 115.15 | 87.71 | 27.44 | |
| *1984* | 118.90 | 86.37 | 32.16 | 0.37 |
| *1985* | 123.58 | 89.90 | 33.24 | 0.44 |
| *1986* | 128.09 | 93.33 | 34.21 | 0.55 |
| *1987* | 132.14 | 96.54 | 34.88 | 0.72 |

**Note:**
*** Figures of staff and workers include 1) staff and workers of state owned units in urban and rural areas; 2) staff and workers of collective owned units in urban areas; 3) staff and workers of other ownership units in urban areas.**

## 2.5 *SOCIAL LABOR FORCE BY SECTOR 1987*

(1,000,000 persons) (year-end figure)

| Sector | Total | Staff and Workers in State Owned Units | Staff and Workers in Collective Owned Units in Cities and Towns | Individual Laborers in Cities and Towns | Collective and Individual Laborers in Rural Areas |
|---|---|---|---|---|---|
| ***National Total*** | **527.83** | **96.54** | **34.88** | **5.69** | **390.00** |
| *Farming, Forestry, Animal Husbandry, Fishery and Water Conservancey* | 317.20 | 8.01 | 0.47 | 0.02 | 308.70 |
| *Industry* | 93.42 | 40.86 | 18.27 | 0.74 | 32.97 |
| *Geological Survey and Prospecting* | 1.07 | 1.07 | | | |
| *Construction* | 24.19 | 6.07 | 3.79 | 0.06 | 14.13 |
| *Transportation, Communication, Postal and Telecommunication Services* | 13.73 | 5.75 | 2.04 | 0.31 | 5.62 |
| *Commerce, Food Service, Supply and Marketing, and Storage of Materials* | 26.56 | 8.94 | 7.60 | 3.91 | 6.07 |
| *Housing Management, Public Utilities, Community Service* | 5.40 | 2.37 | 0.98 | 0.59 | 1.38 |
| *Health Care, Sports and Social Welfare* | 4.96 | 2.96 | 0.69 | 0.04 | 1.27 |
| *Education, Culture, Art, Radio and TV Broadcasting* | 13.75 | 10.25 | 0.34 | 0.02 | 3.14 |
| *Scientific Research and General Technical Service* | 1.58 | 1.40 | 0.02 | | 0.16 |
| *Banking and Insurance* | 1.70 | 1.13 | 0.41 | | 0.16 |
| *Government Agencies, Parties and Social Organizations** | 9.25 | 7.78 | 0.27 | | 1.20 |
| *Others* | 15.02 | | | | 15.02 |

**Note:**
*** Parties and social organizations include Communist Party, other political parties, and social organizations, such as, trade union, women union, youth union, etc.**

## 2.6 *NUMBER OF STAFF AND WORKERS BY SECTOR*

(1,000,000 persons) (year-end figure)

| | Number of Staff and Workers | | Percentage | |
|---|---|---|---|---|
| **Sector** | 1986 | 1987 | 1986 | 1987 |
| ***National Total*** | **128.09** | **132.14** | **100.0** | **100.0** |
| *Farming, Forestry, Animal Husbandry, Fishery and Water Conservancy* | 8.41 | 8.48 | 6.6 | 6.4 |
| *Industry* | 57.81 | 59.71 | 45.1 | 45.2 |
| *Geological Survey and Prospecting* | 1.05 | 1.07 | 0.8 | 0.8 |
| *Construction* | 9.57 | 9.82 | 7.5 | 7.4 |
| *Transportation, Communication, Postal and Telecommunication Service* | 7.75 | 7.80 | 6.0 | 5.9 |
| *Commerce, Food Service, Supply and Marketing, and Storage of Materials* | 16.16 | 16.58 | 12.6 | 12.5 |
| *Housing Management, Public Utilities, Community Service* | 3.27 | 3.43 | 2.5 | 2.6 |
| *Health Care, Sports and Social Welfare* | 3.54 | 3.65 | 2.8 | 2.8 |
| *Education, Culture, Art, Radio and TV Broadcasting* | 10.08 | 10.59 | 7.9 | 8 |
| *Scientific Research and General Technical Service* | 1.37 | 1.42 | 1.1 | 1.1 |
| *Banking and Insurance* | 1.38 | 1.54 | 1.1 | 1.2 |
| *Government Agencies, Parties and Social Organizations* | 7.70 | 8.05 | 6.0 | 6.1 |

## 2.7 *SOCIAL LABOR FORCE BY INDUSTRY*

| | Labor Force (1,000,000 persons) | | | | Percentage | | |
|---|---|---|---|---|---|---|---|
| **Year** | Total | Primary Industries | Secondary Industries | Tertiary Industries | Primary Industries | Secondary Industries | Tertiary Industries |
| *1978* | 398.56 | 294.21 | 60.74 | 43.61 | 73.8 | 15.2 | 11.0 |
| *1980* | 418.96 | 302.06 | 68.21 | 48.69 | 72.1 | 16.3 | 11.6 |
| *1981* | 432.80 | 311.66 | 70.70 | 50.44 | 72.0 | 16.3 | 11.7 |
| *1982* | 447.06 | 320.07 | 72.70 | 54.29 | 71.6 | 16.3 | 12.1 |
| *1983* | 460.04 | 325.04 | 75.04 | 59.96 | 70.7 | 16.3 | 13.0 |
| *1984* | 475.97 | 325.32 | 81.96 | 68.69 | 68.3 | 17.2 | 14.5 |
| *1985* | 498.73 | 311.87 | 105.24 | 81.62 | 62.5 | 21.1 | 16.4 |
| *1986* | 512.82 | 313.11 | 113.57 | 86.14 | 61.1 | 22.1 | 16.8 |
| *1987* | 527.83 | 317.20 | 118.67 | 91.96 | 60.1 | 22.5 | 17.4 |

## 2.8 *NUMBER OF STAFF AND WORKERS BY SECTOR AND REGION 1987*

(1,000,000 persons) (year-end figure)

| Region | Total | Agriculture, Forestry, Animal Husbandry, Fishery and Water Conservancy | Industry | Geological Survey and Prespecting | Construction | Transportation, Postal and Telecommunication Services | Commerce, Food Service, Supply and Marketing and Storage of Materials |
|---|---|---|---|---|---|---|---|
| ***National Total*** | **132.141** | **8.484** | **59.714** | **1.073** | **9.815** | **7.804** | **16.575** |
| *Beijing* | 4.052 | 0.072 | 1.486 | 0.008 | 0.489 | 0.180 | 0.459 |
| *Tianjin* | 2.815 | 0.042 | 1.432 | 0.056 | 0.229 | 0.148 | 0.332 |
| *Hebei* | 6.098 | 0.221 | 2.821 | 0.106 | 0.528 | 0.337 | 0.740 |
| *Shanxi* | 4.095 | 0.084 | 2.036 | 0.018 | 0.374 | 0.230 | 0.474 |
| *Inner Mongolia* | 3.428 | 0.372 | 1.312 | 0.028 | 0.302 | 0.221 | 0.441 |
| *Liaoning* | 9.692 | 0.452 | 5.189 | 0.059 | 0.846 | 0.468 | 1.129 |
| *Jilin* | 4.861 | 0.253 | 2.389 | 0.047 | 0.361 | 0.253 | 0.665 |
| *Heilongjiang* | 8.066 | 1.093 | 3.772 | 0.055 | 0.605 | 0.403 | 0.942 |
| *Shanghai* | 5.054 | 0.168 | 2.741 | 0.004 | 0.287 | 0.303 | 0.582 |
| *Jiangsu* | 8.453 | 0.400 | 4.276 | 0.033 | 0.466 | 0.548 | 1.072 |
| *Zhejiang* | 4.596 | 0.146 | 2.359 | 0.013 | 0.266 | 0.286 | 0.602 |
| *Anhui* | 4.545 | 0.207 | 1.970 | 0.027 | 0.407 | 0.304 | 0.657 |
| *Fujian* | 2.933 | 0.169 | 1.241 | 0.015 | 0.214 | 0.208 | 0.366 |
| *Jiangxi* | 3.653 | 0.387 | 1.604 | 0.036 | 0.208 | 0.198 | 0.434 |
| *Shandong* | 6.846 | 0.167 | 3.413 | 0.067 | 0.378 | 0.338 | 0.915 |
| *Henan* | 6.445 | 0.184 | 2.917 | 0.070 | 0.385 | 0.381 | 1.002 |
| *Hubei* | 6.587 | 0.616 | 2.882 | 0.032 | 0.429 | 0.414 | 0.843 |
| *Hunan* | 5.152 | 0.428 | 2.238 | 0.036 | 0.305 | 0.338 | 0.635 |
| *Guangdong* | 8.205 | 0.964 | 3.119 | 0.039 | 0.478 | 0.563 | 1.266 |
| *Guangxi* | 2.961 | 0.240 | 1.078 | 0.025 | 0.202 | 0.227 | 0.359 |
| *Sichuan* | 8.812 | 0.218 | 3.963 | 0.086 | 0.699 | 0.550 | 1.177 |
| *Guizhou* | 2.119 | 0.071 | 0.848 | 0.023 | 0.222 | 0.128 | 0.227 |
| *Yunnan* | 2.747 | 0.234 | 0.948 | 0.032 | 0.207 | 0.189 | 0.312 |
| *Tibet* | 0.157 | 0.010 | 0.020 | 0.003 | 0.015 | 0.024 | 0.015 |
| *Shaanxi* | 3.578 | 0.092 | 1.620 | 0.028 | 0.343 | 0.199 | 0.393 |
| *Gansu* | 2.151 | 0.111 | 0.941 | 0.033 | 0.209 | 0.136 | 0.216 |
| *Qinghai* | 0.645 | 0.048 | 0.223 | 0.028 | 0.090 | 0.048 | 0.054 |
| *Ningxia* | 0.607 | 0.070 | 0.240 | 0.012 | 0.050 | 0.028 | 0.058 |
| *Xinjiang* | 2.788 | 0.971 | 0.636 | 0.054 | 0.221 | 0.154 | 0.208 |

## NUMBER OF STAFF AND WORKERS BY SECTOR AND REGION 1987

(1,000,000 persons) (year-end figure) (continued)

| Region | Housing Management, Public Utilities, Community Services | Health Care, Sports and Social Welfare | Education, Culture, Arts, Radio and Television Broadcasting | Scientific Research and Technical Services | Banking and Insurance | Goverment Agencies, Parties and Social Organizations |
|---|---|---|---|---|---|---|
| **National Total** | **3.427** | **3.655** | **10.588** | **1.417** | **1.535** | **8.054** |
| Beijing | 0.260 | 0.118 | 0.383 | 0.235 | 0.034 | 0.328 |
| Tianjin | 0.126 | 0.071 | 0.204 | 0.043 | 0.020 | 0.112 |
| Hebei | 0.143 | 0.141 | 0.484 | 0.041 | 0.084 | 0.452 |
| Shanxi | 0.085 | 0.097 | 0.326 | 0.028 | 0.052 | 0.291 |
| Inner Mongolia | 0.070 | 0.087 | 0.276 | 0.027 | 0.046 | 0.246 |
| Liaoning | 0.316 | 0.201 | 0.485 | 0.080 | 0.080 | 0.387 |
| Jilin | 0.128 | 0.120 | 0.339 | 0.035 | 0.051 | 0.220 |
| Heilongjiang | 0.213 | 0.154 | 0.392 | 0.043 | 0.062 | 0.332 |
| Shanghai | 0.249 | 0.142 | 0.317 | 0.104 | 0.029 | 0.128 |
| Jiangsu | 0.210 | 0.212 | 0.742 | 0.075 | 0.078 | 0.341 |
| Zhejiang | 0.118 | 0.137 | 0.341 | 0.029 | 0.061 | 0.238 |
| Anhui | 0.104 | 0.130 | 0.389 | 0.028 | 0.055 | 0.267 |
| Fujian | 0.063 | 0.089 | 0.295 | 0.018 | 0.041 | 0.214 |
| Jiangxi | 0.063 | 0.108 | 0.309 | 0.034 | 0.046 | 0.226 |
| Shandong | 0.138 | 0.238 | 0.601 | 0.045 | 0.088 | 0.458 |
| Henan | 0.129 | 0.214 | 0.530 | 0.050 | 0.094 | 0.489 |
| Hubei | 0.135 | 0.203 | 0.498 | 0.054 | 0.075 | 0.414 |
| Hunan | 0.076 | 0.160 | 0.461 | 0.040 | 0.068 | 0.367 |
| Guangdong | 0.270 | 0.227 | 0.571 | 0.063 | 0.114 | 0.531 |
| Guangxi | 0.074 | 0.094 | 0.352 | 0.030 | 0.046 | 0.234 |
| Sichuan | 0.178 | 0.285 | 0.851 | 0.116 | 0.112 | 0.577 |
| Guizhou | 0.043 | 0.066 | 0.242 | 0.019 | 0.025 | 0.205 |
| Yunnan | 0.044 | 0.085 | 0.338 | 0.030 | 0.046 | 0.282 |
| Tibet | 0.004 | 0.010 | 0.019 | 0.002 | 0.002 | 0.033 |
| Shaanxi | 0.078 | 0.112 | 0.325 | 0.083 | 0.046 | 0.260 |
| Gansu | 0.041 | 0.055 | 0.179 | 0.027 | 0.030 | 0.173 |
| Qinghai | 0.013 | 0.020 | 0.051 | 0.008 | 0.009 | 0.053 |
| Ningxia | 0.016 | 0.016 | 0.059 | 0.001 | 0.009 | 0.042 |
| Xinjiang | 0.040 | 0.063 | 0.229 | 0.023 | 0.032 | 0.157 |

## 2.9 NUMBER OF NEWLY EMPLOYED PERSONS IN URBAN AREAS

(1,000,000 persons)

| Item | 1978 | 1980 | 1984 | 1985 | 1986 | 1987 |
|---|---|---|---|---|---|---|
| ***Total*** | **5.444** | **9.000** | **7.215** | **8.136** | **7.931** | **7.991** |
| ***Major Sources*** | | | | | | |
| *Persons from Cities and Towns* | 2.749 | 6.225 | 4.497 | 5.023 | 4.316 | 4.117 |
| *Persons from Rural Areas* | 1.484 | 1.274 | 1.230 | 1.502 | 1.665 | 1.669 |
| *Graduates from Universities, Colleges and Secondary Technical Schools* | 0.377 | 0.800 | 0.817 | 0.835 | 0.993 | 1.171 |
| *Others* | 0.834 | 0.701 | 0.671 | 0.726 | 0.957 | 1.035 |
| ***Assignment*** | | | | | | |
| *State Owned Units** | 3.920 | 5.722 | 4.156 | 4.991 | 5.363 | 4.994 |
| *Collective Owned Units* | 1.524 | 2.780 | 1.973 | 2.038 | 2.238 | 2.140 |
| *Individual Laborers* | | 0.498 | 1.086 | 1.107 | 0.330 | 0.857 |

**Note:**
*** The 1984, 1985 and 1986 figures include those employed by units wholly or jointly owned by the state.**

## 2.10 NUMBER OF UNEMPLOYED AND PERCENTAGE OF UNEMPLOYMENT IN URBAN AREAS

(1,000,000 persons)

| Year | Unemployed Persons | Youth* | | Unemployment Rate** |
|---|---|---|---|---|
| | | Total Number | Percentage of Total Unemployed Persons | |
| *1978* | 5.300 | 2.491 | 47.0 | 5.3 |
| *1980* | 5.415 | 3.825 | 70.6 | 4.9 |
| *1981* | 4.395 | 3.430 | 78.0 | 3.8 |
| *1982* | 3.794 | 2.938 | 77.4 | 3.2 |
| *1983* | 2.714 | 2.220 | 81.8 | 2.3 |
| *1984* | 2.357 | 1.959 | 83.1 | 1.9 |
| *1985* | 2.385 | 1.969 | 82.6 | 1.8 |
| *1986* | 2.644 | 2.093 | 79.2 | 2.0 |
| *1987* | 2.766 | 2.351 | 85.0 | 2.0 |

**Note:**
*** Includes persons from age 16 to 25.**

**** Urban Unemployment Rate = (Number of Urban Unemployed) / (Total Urban Social Labor Force + Number of Urban Unemployed) x 100%**
**The number of urban unemployed includes those who are aged 16-50 for male and 16-45 for female, ablebodied, registered in the local community and seeking to participate in the labor force. It excludes those who are self employed and have income equal or higher than first grade workers, and those who are not seeking to participate in the labor force.**

*Section III*

# AGRICULTURE

## 3.1 *RURAL GOVERNMENT, POPULATION AND SOCIAL LABOR FORCE BY SECTOR*

| Item | | 1986 | 1987 |
|---|---|---|---|
| **Number of Townships and Towns** | | **71521** | **68490** |
| Township Governments | | 61766 | 58210 |
| *Minority AutonomousGovernments* | | 3746 | 3248 |
| Town Governments | | 9755 | 10280 |
| *Minority Autonomous Governments* | | 113 | 174 |
| Village Committees | | 847894 | 830302 |
| **Number of Households, Population and Labor Force** | | | |
| Number of Households in Rural Areas | 1,000,000 households | **195.745** | **201.683** |
| Total Population in Rural Areas | 1,000,000 persons | 850.072 | 857.131 |
| Social Labor Force in Rural Areas | 1,000,000 persons | 379.898 | 390.004 |
| *Farming, Forestry, Animal Husbandry and Fishery* | 1,000,000 persons | 304.679 | 308.700 |
| *Industry* | 1,000,000 persons | 31.393 | 32.972 |
| *Construction* | 1,000,000 persons | 13.086 | 14.313 |
| *Transportation, Postal and Telecommunications* | 1,000,000 persons | 5.061 | 5.625 |
| *Commerce and Food Service* | 1,000,000 persons | 5.318 | 6.069 |
| *Housing Management, Public Utilities, Community and Consultancy Services* | 1,000,000 persons | 1.262 | 1.381 |
| *Health Care, Sports and Social Welfare* | 1,000,000 persons | 1.246 | 1.270 |
| *Education, Culture, Radio and Television Broadcasting* | 1,000,000 persons | 3.147 | 3.139 |
| *Scientific Research and General Technical Service* | 1,000,000 persons | 0.154 | 0.156 |
| *Banking and Insurance* | 1,000,000 persons | 0.142 | 0.162 |
| *Rural Administrative Organizations* | 1,000,000 persons | 1.034 | 1.196 |
| *Other** | 1,000,000 persons | 13.376 | 15.021 |

**Note:**
*** Including those who are temporarily employed in urban areas.**

## 3.2 *TOTAL OUTPUT VALUE OF SOCIETY IN RURAL AREAS**

(1,000,000,000 yuan)

| Item | 1985 Total | 1985 Percentage | 1986 Total | 1986 Percentage | 1987 Total | 1987 Percentage |
|---|---|---|---|---|---|---|
| ***Total*** | **634.004** | **100.0** | **755.4230** | **100.0** | **943.161** | **100.0** |
| *Gross Output Value of Agriculture* | 361.949 | 57.1 | 401.3010 | 53.1 | 467.570 | 49.6 |
| *Gross Output Value of Rural Industry* | 175.008 | 27.6 | 238.0790 | 31.5 | 328.486 | 34.8 |
| *Gross Output Value of Rural Construction* | 51.049 | 8.1 | 59.1930 | 7.8 | 72.331 | 7.7 |
| *Gross Output Value of Rural Transportation* | 19.042 | 3.0 | 24.5400 | 3.3 | 33.447 | 3.5 |
| *Gross Output Value of Rural Commerce and Food Service* | 26.956 | 4.2 | 32.3100 | 4.3 | 41.327 | 4.4 |

Note:
* Figures in this table are at current price.

## 3.3 *GROSS OUTPUT VALUE OF AGRICULTURE**

( 1,000,000,000 yuan)

| Year | Total Output Value | Farming | Forestry | Animal Husbandry | Sideline Production** | Fishing |
|---|---|---|---|---|---|---|
| *1978* | 139.70 | 107.16 | 4.81 | 20.93 | 4.59 | 2.21 |
| *1979* | 169.76 | 126.73 | 6.07 | 28.56 | 5.80 | 2.60 |
| *1980* | 192.26 | 137.81 | 8.14 | 35.42 | 7.60 | 3.29 |
| *1981* | 218.06 | 153.73 | 9.89 | 40.22 | 9.85 | 4.32 |
| *1982* | 248.33 | 175.03 | 11.01 | 45.58 | 11.59 | 5.12 |
| *1983* | 275.00 | 194.14 | 12.72 | 48.42 | 13.40 | 6.32 |
| *1984* | 321.41 | 219.51 | 16.16 | 58.62 | 18.62 | 8.50 |
| *1985* | 361.95 | 227.98 | 18.87 | 79.69 | 22.80 | 12.61 |
| *1986* | 401.30 | 249.83 | 20.12 | 87.35 | 27.56 | 16.44 |
| *1987* | 467.57 | 283.79 | 22.20 | 106.58 | 32.51 | 22.49 |

Note:
* Figures are at current prices.

** Sideline production includes collection of wild plants, hunting of wild animals and fowls, and handicraft and other industries concurrently run by farm households. Before 1984 village run industry was included in this category. Since 1984 it includes only those businesses run by rural households.

## 3.4 *INDEXES OF GROSS OUTPUT VALUE OF AGRICULTURE*
(1978 = 100)

| Year | Gross Output Value of Agriculture | Farming | Forestry | Animal Husbandry | Sideline Production** | Fishing |
|---|---|---|---|---|---|---|
| *1978* | 100.0 | 100.0 | 100.0 | 100.0 | 100.0 | 100.0 |
| *1979* | 107.6 | 107.2 | 101.4 | 114.6 | 96.5 | 96.6 |
| *1980* | 109.1 | 106.6 | 113.7 | 122.6 | 102.4 | 103.9 |
| *1981* | 116.2 | 112.9 | 118.4 | 129.8 | 127.0 | 108.5 |
| *1982* | 129.3 | 124.5 | 128.5 | 147.0 | 154.8 | 121.8 |
| *1983* | 139.3 | 134.8 | 141.6 | 152.8 | 172.8 | 132.4 |
| *1984* | 156.4 | 148.2 | 168.5 | 173.2 | 229.9 | 155.8 |
| *1985* | 161.8 | 145.3 | 176.1 | 203.0 | 277.3 | 185.1 |
| *1986* | 167.3 | 146.6 | 169.8 | 214.2 | 332.6 | 223.1 |
| *1987* | 177.1 | 154.4 | 169.3 | 221.1 | 383.8 | 263.5 |

**Note:**
*** Figures in this table are at comparable prices.**

**** Sideline production includes collection of wild plants, hunting of wild animals and fowls, and handicraft and other industries concurrently run by farm households. Before 1984 village run industry was included in this category. Since 1984 it includes only those businesses run by rural households.**

## 3.5 *PERCENTAGE OF GROSS OUTPUT VALUE OF AGRICULTURE**
(gross agricultural output value = 100)

| Year | Farming | Forestry | Animal Husbandry | Sideline Production** | Fishing |
|---|---|---|---|---|---|
| *1978* | 76.71 | 3.44 | 14.98 | 3.29 | 1.58 |
| *1979* | 74.65 | 3.58 | 16.82 | 3.42 | 1.53 |
| *1980* | 71.68 | 4.23 | 18.42 | 3.95 | 1.71 |
| *1981* | 70.50 | 4.54 | 18.44 | 4.52 | 2.00 |
| *1982* | 70.48 | 4.43 | 18.35 | 4.67 | 2.06 |
| *1983* | 70.60 | 4.63 | 17.61 | 4.87 | 2.30 |
| *1984* | 68.30 | 5.03 | 18.24 | 5.79 | 2.64 |
| *1985* | 62.99 | 5.21 | 22.02 | 6.30 | 3.48 |
| *1986* | 62.26 | 5.01 | 21.77 | 6.87 | 4.10 |
| *1987* | 60.70 | 4.75 | 22.79 | 6.95 | 4.81 |

**Note:**
*** Figures are at current prices.**

**** Sideline production includes collection of wild plants, hunting of wild animals and fowls, and handicraft and other industries concurrently run by farm households. Before 1984 village run industry was included in this category. Since 1984 it includes only those businesses run by rural households.**

## 3.6 *GROSS OUTPUT VALUE OF AGRICULTURE BY REGION 1987 **

(1,000,000,000 yuan)

| Year | Total Output Value | Farming | Forestry | Animal Husbandry | Sideline Production** | Fishing |
|---|---|---|---|---|---|---|
| ***National Total*** | **467.570** | **283.793** | **22.198** | **106.578** | **32.515** | **22.486** |
| *Beijing* | 3.442 | 1.912 | 0.080 | 1.224 | 0.121 | 0.105 |
| *Tianjin* | 3.293 | 2.024 | 0.037 | 0.722 | 0.223 | 0.287 |
| *Hebei* | 20.066 | 13.780 | 0.683 | 4.363 | 0.811 | 0.429 |
| *Shanxi* | 6.142 | 4.236 | 0.337 | 1.236 | 0.319 | 0.014 |
| *Inner Mongolia* | 8.778 | 4.506 | 0.363 | 2.902 | 0.983 | 0.065 |
| *Liaoning* | 16.922 | 9.610 | 0.486 | 3.958 | 1.226 | 1.642 |
| *Jilin* | 12.081 | 8.612 | 0.293 | 2.340 | 0.708 | 0.128 |
| *Heilongjiang* | 13.700 | 9.975 | 0.669 | 2.282 | 0.520 | 0.254 |
| *Shanghai* | 3.884 | 1.769 | 0.034 | 1.548 | 0.054 | 0.479 |
| *Jiangsu* | 38.025 | 22.811 | 0.602 | 8.745 | 3.029 | 2.338 |
| *Zhejiang* | 22.812 | 11.391 | 1.161 | 4.945 | 2.724 | 2.591 |
| *Anhui* | 25.534 | 17.615 | 1.088 | 4.933 | 1.168 | 0.730 |
| *Fujian* | 13.297 | 6.003 | 1.357 | 2.771 | 1.209 | 1.957 |
| *Jiangxi* | 14.435 | 7.976 | 1.087 | 3.433 | 1.416 | 0.523 |
| *Shandong* | 41.318 | 29.905 | 1.211 | 6.413 | 1.473 | 2.316 |
| *Henan* | 32.362 | 22.974 | 1.324 | 5.111 | 2.762 | 0.191 |
| *Hubei* | 24.968 | 16.013 | 0.997 | 5.486 | 1.053 | 1.419 |
| *Hunan* | 25.339 | 13.831 | 1.814 | 7.240 | 1.396 | 1.058 |
| *Guangdong* | 38.987 | 18.994 | 3.082 | 8.532 | 4.208 | 4.171 |
| *Guangxi* | 13.792 | 7.334 | 0.954 | 3.952 | 1.058 | 0.494 |
| *Sichuan* | 38.894 | 22.722 | 1.773 | 11.848 | 2.015 | 0.536 |
| *Guizhou* | 9.225 | 5.153 | 0.550 | 2.419 | 1.046 | 0.057 |
| *Yunnan* | 11.125 | 6.175 | 0.885 | 2.924 | 1.028 | 0.095 |
| *Tibet* | 1.029 | 0.364 | 0.015 | 0.562 | 0.088 | |
| *Shaanxi* | 10.339 | 6.929 | 0.613 | 1.881 | 0.879 | 0.037 |
| *Gansu* | 6.551 | 4.184 | 0.345 | 1.585 | 0.431 | 0.006 |
| *Qinghai* | 1.594 | 0.703 | 0.066 | 0.684 | 0.139 | 0.002 |
| *Ningxia* | 1.476 | 1.040 | 0.044 | 0.322 | 0.051 | 0.019 |
| *Xinjiang* | 8.164 | 5.252 | 0.248 | 2.199 | 0.422 | 0.043 |

**Note:**
*** Figures in this table are at current prices.**

**** Sideline production includes collection of wild plants, hunting of wild animals and fowls, and handicraft and other industries concurrently run by farm households. Before 1984 village run industry was included in this category. Since 1984 it includes only those businesses run by rural households.**

## 3.7 *SOWN AREA OF SELECTED CROPS*

(1,000,000 mu)*

| Item | 1978 | 1980 | 1986 | 1987 |
|---|---|---|---|---|
| **Total Sown Area** | **2251.561** | **2195.693** | **2163.060** | **2174.348** |
| **Grain** | 1808.805 | 1758.514 | 1663.989 | 1669.020 |
| Rice | 516.313 | 508.177 | 483.992 | 482.892 |
| Wheat | 437.739 | 433.418 | 444.244 | 431.969 |
| Corn | 299.417 | 305.292 | 286.855 | 303.175 |
| Soybeans | 107.156 | 108.395 | 124.418 | 126.674 |
| Tubers | 176.944 | 152.302 | 130.270 | 133.014 |
| **Industrial Crops** | 216.602 | 238.815 | 304.277 | 310.865 |
| Cotton | 72.996 | 73.804 | 64.592 | 72.663 |
| Oil Bearing Crops | 93.335 | 118.927 | 171.217 | 167.707 |
| *Peanuts* | 26.522 | 35.086 | 48.802 | 45.332 |
| *Rapeseed* | 38.995 | 42.661 | 73.737 | 79.011 |
| *Sesame* | 9.565 | 11.642 | 15.099 | 13.041 |
| Hemp Crops | 11.261 | 9.997 | 11.432 | 14.500 |
| *Jute/Ambary Hemp* | 6.182 | 4.711 | 5.174 | 4.082 |
| Sugar Crops | 13.192 | 13.834 | 22.054 | 20.350 |
| *Sugarcane* | 8.228 | 7.193 | 14.249 | 12.884 |
| *Beetroot* | 4.964 | 6.641 | 7.805 | 7.466 |
| Tobacco | 11.759 | 7.679 | 16.880 | 16.923 |
| *Flue-cured Tobacco* | 9.193 | 5.951 | 13.415 | 13.701 |
| Medicinal Crops | 3.242 | 2.011 | 2.694 | 2.378 |
| Other Industrial Crops | 10.817 | 12.563 | 15.408 | 16.344 |
| **Other Farm Crops** | 226.154 | 198.364 | 194.794 | 194.463 |
| Vegetables/Melons | 56.053 | 54.097 | 96.594 | 99.358 |
| Plants Used for Fertilizer | 137.071 | 113.092 | 66.308 | 62.177 |

**Note:**
*** Mu is a Chinese unit for measuring land areas. One acre equals 6.07 mu or one hectare equals 15 mu.**

## 3.8 *TOTAL SOWN AREA OF FARM CROPS*
(1,000,000 mu)*

| Year | Total Sown Area | Grain | | Industrial Crops | |
|---|---|---|---|---|---|
| | | Sown Area | Percentage to Total Sown Area | Sown Area | Percentage to Total Sown Area |
| *1978* | 2251.56 | 1808.81 | 80.3 | 216.60 | 9.6 |
| *1979* | 2227.15 | 1788.94 | 80.3 | 221.51 | 10.0 |
| *1980* | 2195.69 | 1758.51 | 80.1 | 238.82 | 10.9 |
| *1981* | 2177.36 | 1724.37 | 79.2 | 263.41 | 12.1 |
| *1982* | 2171.32 | 1701.94 | 78.4 | 281.91 | 12.3 |
| *1983* | 2159.90 | 1710.71 | 79.2 | 266.41 | 13.0 |
| *1984* | 2163.32 | 1693.26 | 78.3 | 289.32 | 13.4 |
| *1985* | 2154.39 | 1632.68 | 75.8 | 335.67 | 15.6 |
| *1986* | 2163.06 | 1663.98 | 76.9 | 304.28 | 14.1 |
| *1987* | 2174.35 | 1669.02 | 76.8 | 310.87 | 14.3 |

Note:
* Mu is a Chinese unit for measuring land areas. One acre equals 6.07 mu or one hectare equals 15 mu.

## 3.9 *OUTPUT OF MAJOR FARM PRODUCTS*
(1,000,000 tons)

| Year | Grain | | | | Soybeans | Tubers |
|---|---|---|---|---|---|---|
| | Total | Rice | Wheat | Corn | | |
| *1978* | 304.77 | 136.93 | 53.84 | 55.95 | 7.57 | 31.74 |
| *1979* | 332.12 | 143.75 | 62.73 | 60.04 | 7.46 | 28.46 |
| *1980* | 320.56 | 139.91 | 55.21 | 62.60 | 7.94 | 28.73 |
| *1981* | 325.02 | 143.96 | 59.64 | 59.21 | 9.33 | 25.97 |
| *1982* | 354.50 | 161.60 | 68.47 | 60.56 | 9.03 | 27.05 |
| *1983* | 387.28 | 168.87 | 81.39 | 68.21 | 9.76 | 29.25 |
| *1984* | 407.31 | 178.26 | 87.82 | 73.41 | 9.70 | 28.48 |
| *1985* | 379.11 | 168.57 | 85.81 | 63.83 | 10.50 | 26.04 |
| *1986* | 391.51 | 172.22 | 90.04 | 70.86 | 11.61 | 25.34 |
| *1987* | 404.73 | 174.42 | 87.77 | 79.82 | 12.18 | 28.22 |

## OUTPUT OF MAJOR FARM PRODUCTS
(1,000,000 tons) (continued)

| Year | Cotton | Oil Bearing Crops | | | | Jute/Ambary Hemp | Sugarcane | Beetroots | Mulberry Silkworm Cocoons |
|---|---|---|---|---|---|---|---|---|---|
| | | Total | Peanuts | Rapeseed | Sesame | | | | |
| 1978 | 2.167 | 5.218 | 2.377 | 1.868 | 0.322 | 1.088 | 21.116 | 2.702 | 0.173 |
| 1979 | 2.207 | 6.435 | 2.822 | 2.402 | 0.417 | 1.089 | 21.508 | 3.106 | 0.213 |
| 1980 | 2.707 | 7.691 | 3.600 | 2.384 | 0.259 | 1.098 | 22.807 | 6.305 | 0.250 |
| 1981 | 2.968 | 10.255 | 3.826 | 4.065 | 0.510 | 1.260 | 29.668 | 6.360 | 0.252 |
| 1982 | 3.598 | 11.817 | 3.916 | 5.656 | 0.342 | 1.060 | 36.882 | 6.712 | 0.271 |
| 1983 | 4.637 | 10.550 | 3.951 | 4.287 | 0.349 | 1.019 | 31.141 | 9.182 | 0.268 |
| 1984 | 6.258 | 11.910 | 4.815 | 4.205 | 0.476 | 1.492 | 39.519 | 8.284 | 0.306 |
| 1985 | 4.147 | 15.784 | 6.664 | 5.607 | 0.691 | 4.119 | 51.549 | 8.919 | 0.335 |
| 1986 | 3.540 | 14.738 | 5.882 | 5.881 | 0.618 | 1.420 | 50.219 | 8.306 | 0.336 |
| 1987 | 4.245 | 15.278 | 6.171 | 6.605 | 0.526 | 1.137 | 47.864 | 8.140 | 0.354 |

## OUTPUT OF MAJOR FARM PRODUCTS
(1,000,000 tons) (continued)

| Year | Tea | Flue-cured Tobacco | Fruits | | | | | |
|---|---|---|---|---|---|---|---|---|
| | | | Total | Apples | Citrus Fruit | Pears | Grapes | Bananas |
| 1978 | 0.268 | 1.052 | 6.570 | 2.275 | 0.383 | 1.517 | 0.104 | 0.085 |
| 1979 | 0.277 | 0.806 | 7.015 | 2.869 | 0.555 | 1.438 | 0.126 | 0.074 |
| 1980 | 0.304 | 0.717 | 6.793 | 2.363 | 0.713 | 1.466 | 0.110 | 0.061 |
| 1981 | 0.343 | 1.279 | 7.801 | 3.006 | 0.798 | 1.593 | 0.148 | 0.126 |
| 1982 | 0.397 | 1.848 | 7.713 | 2.430 | 0.939 | 1.755 | 0.186 | 0.201 |
| 1983 | 0.401 | 1.151 | 9.487 | 3.541 | 1.296 | 1.795 | 0.247 | 0.207 |
| 1984 | 0.414 | 1.543 | 9.845 | 2.941 | 1.499 | 2.100 | 0.294 | 0.300 |
| 1985 | 0.432 | 2.075 | 11.639 | 3.614 | 1.808 | 2.137 | 0.361 | 0.631 |
| 1986 | 0.461 | 1.374 | 13.477 | 3.337 | 2.548 | 2.348 | 0.442 | 1.251 |
| 1987 | 0.509 | 1.636 | 16.678 | 4.265 | 3.224 | 2.489 | 0.641 | 2.029 |

## 3.10 *NUMBER OF LARGE DOMESTIC ANIMALS*

(1,000,000 head) (year-end figure)

| Year | Large Domestic Animals | | Among Large Domestic Animals | | | | |
|---|---|---|---|---|---|---|---|
| | Total | Draught Animals | Cattle and Buffalo | Horses | Donkeys | Mules | Camels |
| *1978* | 93.89 | 50.23 | 70.72 | 11.25 | 7.48 | 3.87 | 0.57 |
| *1979* | 94.59 | 50.29 | 71.35 | 11.15 | 7.47 | 4.02 | 0.60 |
| *1980* | 95.25 | 50.88 | 71.68 | 11.04 | 7.75 | 4.17 | 0.61 |
| *1981* | 97.64 | 54.71 | 73.30 | 10.97 | 8.42 | 4.33 | 0.63 |
| *1982* | 101.13 | 58.33 | 76.07 | 10.98 | 9.00 | 4.46 | 0.61 |
| *1983* | 103.50 | 61.25 | 78.08 | 10.81 | 9.45 | 4.49 | 0.56 |
| *1984* | 108.39 | 64.03 | 82.13 | 10.98 | 9.96 | 4.79 | 0.53 |
| *1985* | 113.82 | 66.46 | 86.82 | 11.08 | 10.42 | 4.97 | 0.53 |
| *1986* | 118.96 | 69.05 | 91.67 | 10.99 | 10.69 | 5.11 | 0.50 |
| *1987* | 121.91 | 71.13 | 94.65 | 10.69 | 10.85 | 5.25 | 0.48 |

## 3.11 *OUTPUT OF MEAT PRODUCTS*

(year-end figure)

| Year | Output of Pork, Beef and Mutton (1,000,000 tons) | Slaughtered Fatling Pigs* (1,000,000 head) | Pigs (1,000,000 head) | Other Animals (1,000,000 head) | | |
|---|---|---|---|---|---|---|
| | | | | Total | Goats | Sheep |
| *1978* | 8.563 | 161.10 | 301.20 | 169.94 | 73.54 | 96.40 |
| *1979* | 10.624 | 187.68 | 319.71 | 183.14 | 80.57 | 102.57 |
| *1980* | 12.054 | 198.61 | 305.43 | 187.31 | 80.68 | 106.63 |
| *1981* | 12.609 | 194.95 | 293.70 | 187.73 | 78.26 | 109.47 |
| *1982* | 13.508 | 200.63 | 300.73 | 181.79 | 75.22 | 106.57 |
| *1983* | 14.021 | 206.61 | 293.54 | 166.95 | 68.03 | 98.92 |
| *1984* | 15.406 | 220.47 | 306.79 | 158.40 | 63.21 | 95.19 |
| *1985* | 17.607 | 238.75 | 331.40 | 155.88 | 61.67 | 94.21 |
| *1986* | 19.171 | 257.22 | 337.19 | 166.23 | 67.22 | 99.01 |
| *1987* | 19.860 | 261.77 | 327.73 | 180.34 | 77.69 | 102.65 |

**Note:**
*** Including pigs purchased by state and slaughtered by collectives, farmers and other units.**

## 3.12 *OUTPUT OF AQUATIC PRODUCTS*
(1,000,000 tons)

| Year | Total Aquatic Products | Saltwater Aquatic Products | Freshwater Aquatic Products |
|---|---|---|---|
| *1978* | 4.66 | 3.60 | 1.06 |
| *1979* | 4.31 | 3.19 | 1.12 |
| *1980* | 4.50 | 3.26 | 1.24 |
| *1981* | 4.61 | 3.23 | 1.38 |
| *1982* | 5.16 | 3.60 | 1.56 |
| *1983* | 5.46 | 3.62 | 1.84 |
| *1984* | 6.19 | 3.94 | 2.25 |
| *1985* | 7.05 | 4.20 | 2.85 |
| *1986* | 8.24 | 4.76 | 3.48 |
| *1987* | 9.55 | 5.48 | 4.07 |

## 3.13 OUTPUT OF MAJOR AGRICULTURAL PRODUCTS BY REGION 1987

(year-end figure)

| Region | Grain (1,000,000 tons) | Cotton (1,000,000 tons) | Oil Bearing Crops (1,000,000 tons) | Sugar Bearing Crops (1,000,000 tons) | Output of Pork, Beef and Mutton (1,000,000 tons) | Milk (1,000,000 tons) |
|---|---|---|---|---|---|---|
| ***National Total*** | **404.733** | **4.245** | **15.278** | **55.504** | **19.860** | **3.301** |
| *Beijing* | 2.270 | 0.003 | 0.033 | | 0.129 | 0.155 |
| *Tianjin* | 1.677 | 0.013 | 0.052 | | 0.065 | 0.053 |
| *Hebei* | 19.200 | 0.626 | 0.689 | 0.058 | 0.953 | 0.102 |
| *Shanxi* | 7.125 | 0.078 | 0.308 | 0.193 | 0.206 | 0.126 |
| *Inner Mongolia* | 6.070 | | 0.540 | 1.678 | 0.407 | 0.268 |
| *Liaoning* | 12.763 | 0.003 | 0.192 | 0.284 | 0.573 | 0.109 |
| *Jilin* | 16.758 | | 0.353 | 0.690 | 0.343 | 0.082 |
| *Heilongjiang* | 17.376 | | 0.126 | 3.304 | 0.321 | 0.663 |
| *Shanghai* | 2.326 | 0.015 | 0.162 | 0.002 | 0.188 | 0.169 |
| *Jiangsu* | 32.577 | 0.444 | 1.211 | 0.297 | 1.349 | 0.087 |
| *Zhejiang* | 15.890 | 0.065 | 0.397 | 1.088 | 0.762 | 0.133 |
| *Anhui* | 24.287 | 0.186 | 1.511 | 0.115 | 0.769 | 0.024 |
| *Fujian* | 8.393 | | 0.175 | 41.35 | 0.520 | 0.050 |
| *Jiangxi* | 15.628 | 0.059 | 0.357 | 1.908 | 0.781 | 0.022 |
| *Shandong* | 33.937 | 1.244 | 2.343 | 0.038 | 1.264 | 0.050 |
| *Henan* | 29.484 | 0.570 | 1.366 | 0.198 | 0.803 | 0.029 |
| *Hubei* | 23.207 | 0.439 | 0.755 | 0.417 | 1.146 | 0.041 |
| *Hunan* | 25.937 | 0.056 | 0.587 | 1.230 | 1.638 | 0.011 |
| *Guangdong* | 18.482 | | 0.531 | 18.120 | 1.271 | 0.047 |
| *Guangxi* | 12.102 | | 0.186 | 11.674 | 0.711 | 0.008 |
| *Sichuan* | 39.213 | 0.102 | 1.615 | 2.415 | 3.360 | 0.243 |
| *Guizhou* | 6.732 | 0.001 | 0.463 | 0.196 | 0.597 | 0.009 |
| *Yunnan* | 9.348 | | 0.315 | 5.588 | 0.596 | 0.057 |
| *Tibet* | 0.467 | | 0.012 | | 0.088 | 0.141 |
| *Shaanxi* | 9.879 | 0.056 | 0.315 | 0.024 | 0.319 | 0.091 |
| *Gansu* | 5.294 | 0.005 | 0.294 | 0.563 | 0.308 | 0.066 |
| *Qinghai* | 1.041 | | 0.104 | 0.006 | 0.133 | 0.175 |
| *Ningxia* | 1.390 | | 0.044 | 0.317 | 0.047 | 0.023 |
| *Xinjiang* | 5.880 | 0.280 | 0.423 | 0.965 | 0.213 | 0.267 |

## OUTPUT OF MAJOR AGRICULTURAL PRODUCTS BY REGION 1987

(year-end figure) (continued)

| Region | Large Domestic Animals (1,000,000 head) | Slaughtered Fatling Pigs (1,000,000 head) | Pigs (1,000,000 head) | Output of Aquatic Products (1,000,000 tons) |
|---|---|---|---|---|
| ***National Total*** | **121.911** | **261.770** | **327.733** | **9.553** |
| *Beijing* | 0.282 | 1.858 | 1.204 | 0.030 |
| *Tianjin* | 0.274 | 0.797 | 0.505 | 0.081 |
| *Hebei* | 4.923 | 11.634 | 13.636 | 0.179 |
| *Shanxi* | 2.717 | 2.527 | 2.913 | 0.005 |
| *Inner Mongolia* | 7.308 | 2.996 | 4.555 | 0.023 |
| *Liaoning* | 3.153 | 6.264 | 9.304 | 0.806 |
| *Jilin* | 2.745 | 3.632 | 4.161 | 0.045 |
| *Heilongjiang* | 3.146 | 3.716 | 4.384 | 0.102 |
| *Shanghai* | 0.071 | 3.362 | 1.751 | 0.264 |
| *Jiangsu* | 0.857 | 20.192 | 17.433 | 0.922 |
| *Zhejiang* | 0.747 | 12.495 | 13.224 | 1.250 |
| *Anhui* | 5.054 | 9.099 | 11.674 | 0.241 |
| *Fujian* | 1.236 | 6.651 | 8.597 | 0.982 |
| *Jiangxi* | 2.845 | 10.708 | 13.876 | 0.226 |
| *Shandong* | 5.387 | 15.140 | 15.470 | 1.107 |
| *Henan* | 10.008 | 8.552 | 14.047 | 0.076 |
| *Hubei* | 3.471 | 15.344 | 19.255 | 0.583 |
| *Hunan* | 3.802 | 26.574 | 26.478 | 0.444 |
| *Guangdong* | 5.525 | 16.286 | 22.211 | 1.666 |
| *Guangxi* | 6.510 | 8.410 | 15.646 | 0.240 |
| *Sichuan* | 10.265 | 51.114 | 62.027 | 0.172 |
| *Guizhou* | 5.976 | 6.943 | 12.572 | 0.018 |
| *Yunnan* | 9.233 | 7.254 | 18.213 | 0.038 |
| *Tibet* | 5.633 | 0.078 | 0.156 | |
| *Shaanxi* | 2.836 | 4.341 | 6.918 | 0.012 |
| *Gansu* | 5.561 | 4.142 | 5.295 | 0.002 |
| *Qinghai* | 6.025 | 0.566 | 0.894 | 0.002 |
| *Ningxia* | 0.778 | 0.482 | 0.599 | 0.005 |
| *Xinjiang* | 5.543 | 0.613 | 0.735 | 0.014 |

## 3.14 OUTPUT OF MAJOR FORESTRY PRODUCTS
(1,000,000 tons)

| Year | Rubber | Pine Resin | Lacquer | Tong Oil Seed | Tea Oil Seed | Walnuts |
|---|---|---|---|---|---|---|
| 1978 | 0.1016 | 0.3380 | 0.0022 | 0.3910 | 0.4790 | 0.1130 |
| 1979 | 0.1083 | 0.4040 | 0.0027 | 0.3250 | 0.6170 | 0.0910 |
| 1980 | 0.1130 | 0.4210 | 0.0025 | 0.3030 | 0.4900 | 0.1190 |
| 1981 | 0.1277 | 0.5620 | 0.0030 | 0.3600 | 0.6540 | 0.1070 |
| 1982 | 0.1526 | 0.4700 | 0.0029 | 0.3390 | 0.4940 | 0.1030 |
| 1983 | 0.1724 | 0.3040 | 0.0028 | 0.3680 | 0.4350 | 0.1190 |
| 1984 | 0.1888 | 0.3690 | 0.0023 | 0.3620 | 0.5360 | 0.1280 |
| 1985 | 0.1879 | 0.3440 | 0.0022 | 0.3790 | 0.6190 | 0.1220 |
| 1986 | 0.2097 | 0.4160 | 0.0031 | 0.3460 | 0.4380 | 0.1360 |
| 1987 | 0.2376 | 0.5230 | 0.0033 | 0.3420 | 0.5240 | 0.1470 |

## 3.15 PER CAPITA OUTPUT OF MAJOR AGRICULTURAL PRODUCTS
(kg/person)*

| Year | Grain | Cotton | Oil Bearing Crops | Pork, Beef and Mutton | Aquatic Products |
|---|---|---|---|---|---|
| 1978 | 319 | 2.3 | 5.5 | 9.0 | 4.9 |
| 1979 | 343 | 2.3 | 6.7 | 11.0 | 4.5 |
| 1980 | 327 | 2.8 | 7.9 | 12.3 | 4.6 |
| 1981 | 327 | 3.0 | 10.3 | 12.7 | 4.7 |
| 1982 | 352 | 3.6 | 11.7 | 13.4 | 5.1 |
| 1983 | 379 | 4.5 | 10.3 | 13.7 | 5.3 |
| 1984 | 394 | 6.1 | 11.5 | 14.9 | 6.0 |
| 1985 | 363 | 4.0 | 15.1 | 16.9 | 6.6 |
| 1986 | 370 | 3.3 | 13.9 | 18.1 | 7.8 |
| 1987 | 377 | 4.0 | 14.2 | 18.5 | 8.9 |

Note:
* Based on the mean population of the year.

## 3.16 *MAJOR AGRICULTURAL MACHINERY*
(year end figure)

| Year | Total Capacity of Machinery (1,000,000,000 watts) | Number of Large and Medium Size Tractors* | Small Size and Walking Tractors (1,000,000 units) | Large and Medium Size Towed Machinery (1,000,000 units) | Motors for Agricultural Drainage and Irrigation | |
|---|---|---|---|---|---|---|
| | | | | | Total Number (1,000,000 units) | Capacity (1,000,000,000 watts) |
| *1978* | 117.50 | 557358 | 1.373 | 1.192 | 5.026 | 48.23 |
| *1979* | 133.79 | 666823 | 1.671 | 1.313 | 5.384 | 52.38 |
| *1980* | 147.46 | 744865 | 1.874 | 1.369 | 5.630 | 54.90 |
| *1981* | 156.80 | 792032 | 2.037 | 1.390 | 5.672 | 55.15 |
| *1982* | 166.14 | 812447 | 2.287 | 1.374 | 5.803 | 56.41 |
| *1983* | 180.22 | 840776 | 2.750 | 1.308 | 6.077 | 57.73 |
| *1984* | 194.97 | 853914 | 3.298 | 1.235 | 6.150 | 57.77 |
| *1985* | 209.12 | 852357 | 3.824 | 1.128 | 6.163 | 57.55 |
| *1986* | 229.50 | 866463 | 4.526 | 1.006 | 6.507 | 60.44 |
| *1987* | 248.45 | 880952 | 5.300 | 1.035 | 6.839 | 62.58 |

Note:
* Tractors with a capacity of 14.71 kilowatts or more.

### *MAJOR AGRICULTURAL MACHINERY*
(year end figure) (continued)

| Year | Number of Combine Harvesters | Number of Trucks | Animal Drawn Carts with Rubber Tires (1,000,000 units) | Hand Carts with Rubber Tires (1,000,000 units) | Motorized Fishing Boats | |
|---|---|---|---|---|---|---|
| | | | | | Total Number | Capacity (1,000,000,000 watts) |
| *1978* | 18987 | 73770 | 2.488 | 29.634 | 47176 | 2.14 |
| *1979* | 23026 | 97105 | 2.477 | 32.624 | 52225 | 2.30 |
| *1980* | 27045 | 137668 | 2.398 | 35.170 | 61022 | 2.58 |
| *1981* | 31268 | 175126 | 2.337 | 41.260 | 73586 | 2.87 |
| *1982* | 33904 | 206383 | 2.344 | 48.415 | 95692 | 3.22 |
| *1983* | 35728 | 274751 | 2.586 | 55.634 | 120167 | 3.27 |
| *1984* | 35861 | 349261 | 2.852 | 59.450 | 143430 | 3.35 |
| *1985* | 34573 | 429554 | 2.875 | 61.144 | 172582 | 3.67 |
| *1986* | 30945 | 499164 | 3.051 | 62.413 | 205923 | 4.24 |
| *1987* | 33802 | 550192 | 3.482 | 65.107 | 247566 | 4.94 |

## 3.17 AREAS UNDER MECHANIZED PLOUGHING AND IRRIGATION

(1,000,000 hectare)

| Year | Areas Under Mechanized Ploughing | Areas Under Irrigation: Total Area | Areas Under Irrigation: Mechanized Irrigated Areas | Percentage of Mechanized Irrigated Areas |
|---|---|---|---|---|
| 1978 | 40.670 | 44.965 | 24.895 | 55.4 |
| 1979 | 42.219 | 45.003 | 25.321 | 56.3 |
| 1980 | 40.990 | 44.888 | 25.315 | 56.4 |
| 1981 | 36.477 | 44.574 | 25.231 | 56.6 |
| 1982 | 35.115 | 44.177 | 25.145 | 56.9 |
| 1983 | 33.572 | 44.644 | 25.265 | 56.6 |
| 1984 | 34.922 | 44.453 | 25.062 | 56.4 |
| 1985 | 34.442 | 44.036 | 24.629 | 55.9 |
| 1986 | 36.428 | 44.226 | 25.031 | 56.6 |
| 1987 | 38.393 | 44.403 | 24.825 | 55.9 |

## 3.18 CHEMICAL FERTILIZER APPLICATION, NUMBER OF SMALL HYDROELECTRIC POWER STATIONS AND RURAL ELECTRICITY CONSUMPTION

| Year | Application of Chemical Fertilizer* (1,000,000 tons) | Small Rural Hydroelectric Power Stations: Number | Small Rural Hydroelectric Power Stations: Capacity (1,000,000 Kw) | Electricity Consumed in Rural Areas** (1,000,000,000 kwh) |
|---|---|---|---|---|
| 1978 | 8.840 | 82387 | 2.234 | 25.31 |
| 1979 | 10.863 | 83224 | 2.763 | 28.27 |
| 1980 | 12.694 | 80319 | 3.041 | 32.08 |
| 1981 | 13.349 | 74017 | 3.360 | 36.99 |
| 1982 | 15.134 | 66256 | 3.530 | 39.69 |
| 1983 | 16.598 | 62328 | 3.463 | 42.81 |
| 1984 | 17.398 | 60062 | 3.615 | 46.40 |
| 1985 | 17.758 | 55754 | 3.802 | 50.89 |
| 1986 | 19.306 | 54136 | 3.879 | 58.67 |
| 1987 | 19.970 | 51978 | 3.941 | 65.88 |

**Note:**
*** Chemical fertilizer is calculated on the basis of effective ingredients.**

**** Electricity consumption includes electricity provided by state power grid and by rural power stations; it excludes electricity consumed by state owned units in rural areas.**

## 3.19 *RURAL TOWNSHIP AND VILLAGE RUN ENTERPRISES*

| Item | | 1980 | 1985 | 1986 | 1987 |
|---|---|---|---|---|---|
| **Number of Enterprises** | 1,000,000 | **1.425** | **1.569** | **1.510** | **1.579** |
| **Number of Persons Employed** | 1,000,000 | 29.997 | 41.521 | 43.915 | 47.025 |
| Township Enterprises | 1,000,000 | 13.938 | 21.114 | 22.749 | 23.844 |
| Village Enterprises | 1,000,000 | 16.059 | 20.407 | 21.166 | 23.181 |
| **Number of Persons Employed by Sector** | | | | | |
| Agricultural Enterprises | 1,000,000 | 4.561 | 2.524 | 2.408 | 2.441 |
| Industrial Enterprises | 1,000,000 | 19.423 | 27.821 | 30.413 | 33.390 |
| Transport and Telecommunications | 1,000,000 | 1.135 | 1.142 | 1.098 | 1.066 |
| Construction | 1,000,000 | 3.347 | 7.897 | 8.069 | 8.284 |
| Other | 1,000,000 | 1.531 | 2.135 | 1.927 | 1.844 |
| *Commerce* | 1,000,000 | 0.000 | 0.807 | 0.776 | 0.818 |
| *Food Services* | 1,000,000 | 0.000 | 0.273 | 0.224 | 0.230 |
| *Service Trade* | 1,000,000 | 0.000 | 0.621 | 0.418 | 0.420 |
| **Total Revenue** | (1,000,000,000 yuan) | 59.61 | 182.74 | 222.36 | 302.08 |

## 3.20 *TOTAL AREA COVERED AND AFFECTED BY NATURAL DISASTERS*

(1,000,000 mu*)

| Year | Area Covered by Natural Disasters** | Area Affected by Natural Disasters** | | Flood | | Drought | |
|---|---|---|---|---|---|---|---|
| | | Total | Percentage of Covered Areas | Area Covered by Disasters | Area Affected by Disasters | Area Covered by Disasters | Area Affected by Disasters |
| *1978* | 761.85 | 327.00 | 42.9 | 42.75 | 13.80 | 602.55 | 269.55 |
| *1979* | 590.55 | 226.80 | 38.4 | 101.40 | 43.05 | 369.75 | 139.80 |
| *1980* | 667.89 | 334.76 | 50.1 | 137.19 | 75.38 | 391.67 | 187.28 |
| *1981* | 596.79 | 281.15 | 47.1 | 129.37 | 59.59 | 385.39 | 182.01 |
| *1982* | 497.00 | 241.76 | 48.6 | 125.41 | 66.95 | 310.46 | 149.58 |
| *1983* | 520.70 | 243.14 | 46.7 | 182.43 | 86.21 | 241.33 | 113.79 |
| *1984* | 478.31 | 228.96 | 47.9 | 159.48 | 80.92 | 237.28 | 105.22 |
| *1985* | 665.48 | 340.58 | 51.2 | 212.96 | 134.24 | 344.84 | 150.94 |
| *1986* | 707.03 | 354.84 | 50.2 | 137.33 | 84.02 | 465.63 | 221.47 |
| *1987* | 631.29 | 305.89 | 48.5 | 130.29 | 61.56 | 373.80 | 195.49 |

**Note:**
*** Mu is a Chinese unit for measuring land areas. One acre equals 6.07 mu or one hectare equals 15 mu.**

**** Area covered by disasters is measured according to meteorological conditions, e.g., an area is covered by drought if precipitation in this area is less than certain standard. Area affected by disasters, for example, drought, may be smaller than area covered, due to irrigation, etc.**

Section IV

# INDUSTRY

## 4.1 NUMBER OF INDUSTRIAL ENTERPRISES AND GROSS OUTPUT VALUE OF INDUSTRY*

| | | 1985 | | 1986 | | 1987 | |
|---|---|---|---|---|---|---|---|
| | | Total | Township or Above | Total | Township or Above | Total | Township or Above |
| **Industrial Enterprises** | 1,000,000 | 5.185 | 0.463 | 6.706 | 0.499 | 7.533 | 0.494 |
| **Industrial Enterprises by Ownership** | | | | | | | |
| State Owned | 1,000,000 | 0.093 | 0.094 | 0.097 | 0.097 | 0.098 | 0.098 |
| Collective Owned | 1,000,000 | 1.742 | 0.368 | 1.822 | 0.400 | 1.851 | 0.392 |
| *Township Enterprises* | 1,000,000 | 0.217 | 0.217 | 0.246 | 0.246 | 0.238 | 0.238 |
| Individual Owned | 1,000,000 | 3.348 | | 4.782 | | 5.580 | |
| Others** | 1,000,000 | 0.002 | 0.002 | 0.002 | 0.002 | 0.004 | 0.004 |
| **Industrial Enterprises by Type** | | | | | | | |
| Light Industry*** | 1,000,000 | 3.661 | 0.263 | 4.457 | 0.282 | 5.112 | 0.275 |
| Heavy Industry*** | 1,000,000 | 1.524 | 0.200 | 2.159 | 0.217 | 2.421 | 0.219 |
| **Industrial Enterprises by Size** | | | | | | | |
| Large | unit | | 2321 | | 2612 | | 2908 |
| Medium | unit | | 5590 | | 6177 | | 6957 |
| Small | unit | | 4455299 | | 490495 | | 483708 |
| **Gross Output Value of Industry****** | 1,000,000,000 yuan | 971.6 | 872.2 | 1119.4 | 979.9 | 1380.6 | 1182.9 |
| **Gross Output Value of Industry by Ownership** | | | | | | | |
| State Owned | 1,000,000,000 yuan | 630.2 | 630.2 | 697.1 | 697.1 | 825.0 | 825.0 |
| Collective Owned | 1,000,000,000 yuan | 311.7 | 230.3 | 375.2 | 266.5 | 477.7 | 330.0 |
| *Township Enterprises* | 1,000,000,000 yuan | 76.1 | 76.1 | 98.1 | 98.1 | 128.4 | 128.4 |
| Individual Owned | 1,000,000,000 yuan | 18.0 | | 30.8 | | 50.0 | |
| Others** | 1,000,000,000 yuan | 11.7 | 11.7 | 16.3 | 16.3 | 27.9 | 27.9 |
| **Gross Output Value of Industry by Light or Heavy Industry** | | | | | | | |
| Light Industry | 1,000,000,000 yuan | 457.5 | 402.5 | 533.0 | 455.7 | 663.0 | 553.4 |
| Heavy Industry | 1,000,000,000 yuan | 514.1 | 469.7 | 586.4 | 524.2 | 717.6 | 629.5 |
| **Gross Output Value of Industry by Enterprise Size** | | | | | | | |
| Large | 1,000,000,000 yuan | | 245.8 | | 285.2 | | 353.2 |
| Medium | 1,000,000,000 yuan | | 165.2 | | 185.2 | | 224.3 |
| Small | 1,000,000,000 yuan | | 461.2 | | 509.5 | | 605.4 |

**Note:**

*** Including enterprises of all types of ownership with independent and non-independent accounting system but excluding enterprises run by villages and handicraft industries run by farm households.**

**** Including joint state-collective enterprises, joint state-private and joint collective-private enterprises, foreign -Chinese joint ventures, enterprises run by overseas Chinese, and businessmen from Hong Kong and Macao, and enterprises run by foreigners.**

***** Light industry produces comsumer goods and hand tools; heavy industry produces producer goods.**

****** Gross output value of industry are at currrent prices.**

## 4.2 *GROSS OUTPUT VALUE OF INDUSTRY BY REGION 1987**

( 1,000,000,000 yuan)

| Region | Total | State Owned Industry | Collective Owned Industry | | | Individual Owned Industry | Other Industry |
|---|---|---|---|---|---|---|---|
| | | | *Total* | *Township Industry* | *Village Industry* | | |
| ***National Total*** | **1380.64** | **825.01** | **477.77** | **128.42** | **115.60** | **49.98** | **27.88** |
| *Beijing* | 43.32 | 30.60 | 11.34 | 2.88 | 3.15 | 0.18 | 1.20 |
| *Tianjin* | 40.72 | 28.15 | 11.22 | 1.61 | 4.33 | 0.08 | 1.27 |
| *Hebei* | 64.09 | 34.06 | 24.85 | 4.33 | 6.42 | 4.98 | 0.20 |
| *Shanxi* | 30.43 | 19.76 | 9.38 | 2.17 | 3.19 | 1.20 | 0.03 |
| *Inner Mongolia* | 15.11 | 11.98 | 2.79 | 0.36 | 0.25 | 0.33 | 0.01 |
| *Liaoning* | 104.41 | 66.82 | 29.74 | 5.17 | 7.14 | 4.60 | 3.25 |
| *Jilin* | 35.94 | 25.80 | 8.49 | 1.15 | 1.15 | 1.60 | 0.05 |
| *Heilongjiang* | 56.40 | 45.43 | 9.90 | 1.56 | 1.34 | 1.00 | 0.07 |
| *Shanghai* | 106.73 | 79.03 | 21.19 | 5.68 | 5.27 | 0.15 | 6.36 |
| *Jiangsu* | 158.93 | 58.25 | 93.23 | 35.41 | 26.80 | 3.96 | 3.49 |
| *Zhejiang* | 85.07 | 28.27 | 53.17 | 18.93 | 12.64 | 2.75 | 0.88 |
| *Anhui* | 39.41 | 24.21 | 12.13 | 3.17 | 1.67 | 2.96 | 0.11 |
| *Fujian* | 26.39 | 13.96 | 9.09 | 1.64 | 2.16 | 1.04 | 2.30 |
| *Jiangxi* | 25.85 | 18.41 | 6.33 | 1.79 | 1.25 | 1.00 | 0.11 |
| *Shandong* | 103.29 | 52.17 | 47.02 | 11.81 | 14.77 | 3.90 | 0.20 |
| *Henan* | 59.67 | 35.11 | 19.29 | 2.88 | 5.38 | 5.18 | 0.09 |
| *Hubei* | 65.91 | 43.49 | 20.14 | 4.61 | 3.78 | 2.11 | 0.17 |
| *Hunan* | 45.67 | 29.77 | 13.70 | 3.98 | 2.70 | 2.12 | 0.08 |
| *Guangdong* | 89.09 | 44.27 | 34.21 | 8.49 | 6.61 | 3.45 | 7.16 |
| *Guangxi* | 20.75 | 15.46 | 4.28 | 0.79 | 0.36 | 0.94 | 0.07 |
| *Sichuan* | 72.50 | 48.94 | 19.96 | 6.32 | 2.60 | 3.27 | 0.33 |
| *Guizhou* | 12.67 | 10.00 | 1.68 | 0.45 | 0.12 | 0.83 | 0.16 |
| *Yunnan* | 18.19 | 14.02 | 3.64 | 1.06 | 0.26 | 0.43 | 0.08 |
| *Tibet* | 0.21 | 0.15 | 0.04 | 0.02 | 0.00 | 0.00 | 0.00 |
| *Shaanxi* | 25.86 | 18.53 | 6.11 | 1.17 | 1.74 | 1.16 | 0.06 |
| *Gansu* | 15.99 | 13.50 | 0.50 | 2.15 | 0.59 | 0.31 | 0.29 |
| *Qinghai* | 3.12 | 2.61 | 0.46 | 0.05 | 0.05 | 0.05 | 0.00 |
| *Ningxia* | 3.34 | 2.60 | 0.61 | 0.14 | 0.04 | 0.12 | 0.01 |
| *Xinjiang* | 11.60 | 9.61 | 1.63 | 0.23 | 0.12 | 0.22 | 0.14 |

**Note:**
*** Figures in this table are at the current prices.**

# 4.3 NUMBER OF ENTERPRISES AND GROSS OUTPUT VALUE OF SELECTED INDUSTRIES*

| Industries | 1986 | | 1987 | | Gross Output Value in 1987 as Percentage of 1986 |
|---|---|---|---|---|---|
| | Number of Enterprises | Gross Output Value (1,000,000,000 yuan) | Number of Enterprises | Gross Output Value (1,000,000,000 yuan) | |
| *Coal Mining and Preparation* | 10962 | 23.744 | 10553 | 25.698 | 103.4 |
| *Petroluem and Natural Gas Extraction* | 32 | 23.081 | 98 | 29.143 | 102.4 |
| *Ferrous Metal Mining* | 1517 | 1.995 | 1426 | 2.309 | 113.8 |
| *Non-Ferrous Metal Mining* | 2300 | 4.563 | 2438 | 5.796 | 115.9 |
| *Construction Material and Non-Metal Mineral Preparation* | 11868 | 4.404 | 11347 | 5.183 | 114.1 |
| *Food Manufacturing* | 65556 | 80.600 | 61655 | 92.502 | 106.7 |
| *Beverage Manufacturing* | 18976 | 17.986 | 18537 | 24.302 | 123.4 |
| *Tobacco Manufatcuring* | 407 | 22.588 | 413 | 27.997 | 120.3 |
| *Textile Industry* | 23306 | 119.854 | 25505 | 141.066 | 110.6 |
| *Sewing Industry* | 23551 | 19.437 | 21614 | 23.590 | 116.0 |
| *Leather, Furs and Related Products* | 8505 | 10.203 | 8883 | 12.252 | 116.5 |
| *Timber Processing and Bamboo, Rattan, Palm and Straw Products* | 14832 | 7.402 | 13574 | 9.264 | 110.2 |
| *Paper Making and Manufactured Goods* | 10008 | 18.090 | 10812 | 23.400 | 117.9 |
| *Power Generation, Steam and Hot Water Production and Supply* | 13948 | 31.323 | 13896 | 37.089 | 110.5 |
| *Petroleum Processing* | 577 | 29.858 | 736 | 34.233 | 110.9 |
| *Chemical Industry* | 16558 | 64.644 | 17980 | 83.584 | 117.3 |
| *Medical and Pharmaceutical Products* | 2966 | 15.725 | 3150 | 21.148 | 125.6 |
| *Chemical Fibers* | 408 | 10.156 | 467 | 12.982 | 125.6 |
| *Rubber Manufactured Goods* | 3789 | 14.958 | 3993 | 16.776 | 110.1 |
| *Plastic Manufactured Goods* | 14487 | 11.769 | 15336 | 21.563 | 116.8 |
| *Building Materials and Other Non-Metal Mineral Products* | 63543 | 54.533 | 63674 | 62.953 | 112.6 |
| *Ferrous Metals Smelting and Pressing* | 3020 | 65.893 | 2946 | 77.012 | 109.4 |
| *Non-Ferrous Metal Smelting and Pressing* | 1784 | 23.703 | 2110 | 27.829 | 108.8 |
| *Metal Products* | 33625 | 28.665 | 34223 | 35.159 | 120.2 |
| *Engineering Industry* | 48094 | 103.577 | 46957 | 125.869 | 118.1 |
| *Transportation Equipment Manufatcuring* | 11786 | 36.085 | 1264 | 44.189 | 121.7 |
| *Electric Equipment and Machinery* | 14147 | 41.240 | 15861 | 48.812 | 115.7 |
| *Electric and Telecommunication Equipment* | 4330 | 25.334 | 4524 | 35.828 | 144.5 |
| *Precision and Metrological Instruments* | 3639 | 7.322 | 3926 | 8.376 | 113.1 |

**Note:**
*** Figures of gross output value are at current prices. Indexes are based on comparable prices.**

## 4.4 *OUTPUT OF MAJOR INDUSTRIAL PRODUCTS*

| Year | Chemical Fibers (1,000,000 tn) Total | Chemical Fibers (1,000,000 tn) Synthetic Fibers | Yarn* (1,000,000 tn) | Cloth (1,000,000,000 m) | Woolen Fabric (1,000,000 m) | Silk (1,000,000 tn) | Silk Textiles (1,000,000,000 m) | Gunny Bags (1,000,000,000) | Machine-made Paper, Paperboard (1,000,000 tn) |
|---|---|---|---|---|---|---|---|---|---|
| *1978* | 0.2846 | 0.1694 | 2.382 | 11.03 | 88.85 | 0.0297 | 0.611 | 0.290 | 4.39 |
| *1979* | 0.3263 | 0.2136 | 2.635 | 12.15 | 90.17 | 0.0297 | 0.663 | 0.344 | 4.93 |
| *1980* | 0.4503 | 0.3141 | 2.926 | 13.47 | 100.95 | 0.0354 | 0.759 | 0.410 | 5.35 |
| *1981* | 0.5273 | 0.3847 | 3.170 | 14.27 | 113.08 | 0.0374 | 0.835 | 0.429 | 5.40 |
| *1982* | 0.5170 | 0.3753 | 3.354 | 15.35 | 126.69 | 0.0371 | 0.914 | 0.500 | 5.89 |
| *1983* | 0.5407 | 0.4020 | 3.270 | 14.88 | 142.91 | 0.0369 | 0.999 | 0.551 | 6.61 |
| *1984* | 0.7349 | 0.5758 | 3.219 | 13.70 | 180.49 | 0.0376 | 1.178 | 0.548 | 7.56 |
| *1985* | 0.9478 | 0.7706 | 3.535 | 14.67 | 218.16 | 0.0422 | 1.449 | 0.627 | 9.11 |
| *1986* | 1.0173 | 0.8307 | 3.978 | 16.47 | 251.87 | 0.0472 | 1.499 | 0.760 | 9.99 |
| *1987* | 1.1569 | 0.9462 | 4.324 | 16.71 | 259.54 | 0.0508 | 1.506 | 0.840 | 10.08 |

**Note:**
*** Includes blend fabrics and yarns, but excludes substitute fiber yarn and cloth, hand-woven yarn and cloth and cord fabrics.**

## *OUTPUT OF MAJOR INDUSTRIAL PRODUCTS*
(continued)

| Year | Sugar (1,000,000 tn) | Beer (1,000,000 tn) | Cigarettes (1,000,000 cases) | Raw Salt (1,000,000 tn) | Chemical Pharma ceuticals (1,000,000 tn) | Synthetic Detergents (1,000,000 tn) | Bicycles (1,000,000 units) | Sewing Machines (1,000,000 units) | Wrist Watches (1,000,000 units) |
|---|---|---|---|---|---|---|---|---|---|
| *1978* | 2.27 | | 11.82 | 19.53 | | 0.324 | 8.540 | 4.865 | 13.511 |
| *1979* | 2.50 | | 13.03 | 14.77 | | 0.397 | 10.095 | 5.868 | 17.070 |
| *1980* | 2.57 | 0.69 | 15.20 | 17.28 | | 0.393 | 13.024 | 7.678 | 22.155 |
| *1981* | 3.17 | 0.91 | 17.04 | 18.32 | | 0.478 | 17.543 | 10.391 | 28.724 |
| *1982* | 3.38 | 1.17 | 18.85 | 16.38 | | 0.569 | 24.200 | 12.860 | 33.010 |
| *1983* | 3.77 | 1.63 | 19.38 | 16.13 | | 0.677 | 27.582 | 10.872 | 34.690 |
| *1984* | 3.80 | 2.24 | 21.32 | 16.42 | | 0.810 | 28.614 | 9.349 | 37.982 |
| *1985* | 4.51 | 3.10 | 23.70 | 14.79 | 0.1577 | 1.005 | 32.277 | 9.912 | 54.311 |
| *1986* | 5.25 | 4.13 | 25.96 | 17.61 | 0.1514 | 1.175 | 35.683 | 9.894 | 73.174 |
| *1987* | 5.11 | 5.25 | 28.81 | 17.69 | 0.1542 | 1.179 | 40.911 | 9.596 | 61.424 |

## OUTPUT OF MAJOR INDUSTRIAL PRODUCTS
(continued)

| Year | Television Sets (1,000,000 units) | | Radios (1,000,000 units) | Cameras (1,000,000 units) | Household Washing Machines (1,000,000 units) | Refrigerators (1,000,000 units) | Electric Fans (1,000,000 units) | Electric Light Bulbs (1,000,000,000 units) | Coal (1,000,000,000 tn) |
|---|---|---|---|---|---|---|---|---|---|
| | Total | Color | | | | | | | |
| *1978* | 0.5173 | 0.0038 | 11.677 | 0.1789 | 0.0004 | | 1.38 | 0.759 | 0.618 |
| *1979* | 1.3285 | 0.0095 | 13.807 | 0.2381 | 0.0180 | | 2.33 | 0.850 | 0.635 |
| *1980* | 2.4920 | 0.0321 | 30.038 | 0.3728 | 0.2450 | 0.0490 | 7.24 | 0.946 | 0.620 |
| *1981* | 5.3941 | 0.1521 | 40.572 | 0.6230 | 1.2810 | 0.0556 | 10.50 | 0.966 | 0.622 |
| *1982* | 5.9201 | 0.2881 | 17.239 | 0.7423 | 2.5330 | 0.0999 | 9.19 | 1.073 | 0.666 |
| *1983* | 6.8401 | 0.5311 | 19.989 | 0.9256 | 3.6590 | 0.1885 | 10.46 | 1.249 | 0.715 |
| *1984* | 10.0381 | 1.3395 | 22.203 | 1.2618 | 5.7810 | 0.5474 | 17.71 | 1.412 | 0.789 |
| *1985* | 16.6766 | 4.3528 | 16.003 | 1.7897 | 8.8720 | 1.4481 | 31.75 | 1.533 | 0.872 |
| *1986* | 14.5940 | 4.1460 | 15.895 | 2.0254 | 8.9340 | 2.2552 | 35.29 | 1.607 | 0.894 |
| *1987* | 19.3784 | 6.7192 | 16.148 | 2.3942 | 9.9152 | 3.9786 | 35.07 | 1.704 | 0.928 |

## OUTPUT OF MAJOR INDUSTRIAL PRODUCTS
(continued)

| Year | Crude Oil (1,000,000 tn) | Natural Gas (1,000,000,000 cu. m) | Electricity (1,000,000,000 kwh) | | Pig Iron (1,000,000 tn) | Steel (1,000,000 tn) | Rolled Steel Final Products* (1,000,000 tn) | Machine Made Coke (1,000,000 tn) | Timber** (1,000,000 cu. m) |
|---|---|---|---|---|---|---|---|---|---|
| | | | Total | Hydro-Electric Power | | | | | |
| *1978* | 104.05 | 13.73 | 256.60 | 44.60 | 34.79 | 31.78 | 22.08 | 32.69 | 51.62 |
| *1979* | 106.15 | 14.51 | 282.00 | 50.10 | 36.73 | 34.48 | 24.97 | 33.54 | 54.39 |
| *1980* | 105.95 | 14.27 | 300.60 | 58.20 | 38.02 | 37.12 | 27.16 | 34.05 | 53.59 |
| *1981* | 101.22 | 12.74 | 309.30 | 65.50 | 34.17 | 35.60 | 36.70 | 31.72 | 49.42 |
| *1982* | 102.12 | 11.93 | 327.70 | 74.40 | 35.51 | 37.16 | 29.02 | 33.11 | 50.41 |
| *1983* | 106.07 | 12.21 | 351.40 | 86.40 | 37.38 | 40.02 | 30.72 | 34.51 | 52.32 |
| *1984* | 114.61 | 12.43 | 377.00 | 86.80 | 40.01 | 43.47 | 33.72 | 36.15 | 63.85 |
| *1985* | 124.90 | 12.93 | 410.70 | 92.40 | 43.84 | 46.79 | 36.93 | 38.36 | 63.23 |
| *1986* | 130.69 | 13.38 | 449.50 | 94.70 | 50.64 | 52.20 | 40.58 | 40.93 | 65.02 |
| *1987* | 134.04 | 13.81 | 497.30 | 99.50 | 54.32 | 56.02 | 43.91 | 44.75 | 68.43 |

**Note:**
*** Rolled steel final products exclude the intermediate steel products used within steel industry.**

**** Timber includes only timber already transported for use. Since 1984, statistics for timber include a wider range of items than previous years.**

## OUTPUT OF MAJOR INDUSTRIAL PRODUCTS
(continued)

| Year | Cement (1,000,000 tn) | Plate Glass (1,000,000 standard boxes) | Sulphuric Acid (1,000,000 tn) | Soda Ash (1,000,000 tn) | Caustic Soda (1,000,000 tn) | Synthetic Ammonia (1,000,000 tn) | Chemical Fertilizers for Agricultural Use* (1,000,000 tn) | | |
|---|---|---|---|---|---|---|---|---|---|
| | | | | | | | Total | Nitrogenous | Phosphate |
| 1978 | 65.24 | 17.84 | 6.610 | 1.329 | 1.640 | 11.835 | 8.693 | 7.639 | 1.033 |
| 1979 | 73.90 | 20.83 | 6.998 | 1.486 | 1.826 | 13.482 | 10.654 | 8.821 | 1.817 |
| 1980 | 79.86 | 24.66 | 7.643 | 1.613 | 1.923 | 14.974 | 12.321 | 9.993 | 2.308 |
| 1981 | 82.90 | 27.01 | 7.807 | 1.652 | 1.923 | 14.833 | 12.390 | 9.857 | 2.508 |
| 1982 | 95.20 | 31.54 | 8.175 | 1.735 | 2.073 | 15.463 | 12.781 | 10.219 | 2.537 |
| 1983 | 108.25 | 36.47 | 8.696 | 1.793 | 2.123 | 16.771 | 13.789 | 11.094 | 2.666 |
| 1984 | 123.02 | 41.90 | 8.172 | 1.880 | 2.222 | 18.374 | 14.602 | 12.210 | 2.360 |
| 1985 | 145.95 | 49.42 | 6.764 | 2.011 | 2.353 | 17.188 | 13.222 | 11.438 | 1.760 |
| 1986 | 166.06 | 52.02 | 7.631 | 2.154 | 2.520 | 16.764 | 13.966 | 11.597 | 2.345 |
| 1987 | 181.28 | 56.02 | 9.620 | 2.372 | 2.735 | 19.487 | 17.033 | 13.599 | 3.384 |

Note:
* Measured in effective ingredients.

## OUTPUT OF MAJOR INDUSTRIAL PRODUCTS
(continued)

| Year | Chemical Pesticides (1,000,000 tn) | Ethylene (1,000,000 tn) | Plastics (1,000,000 tn) | Calcium Carbide (1,000,000 tn) | Tires (1,000,000 units) | Mining Equipment (1,000,000 tn) | Power Generating Equipment* (1,000,000 Kw) | Metal Cutting Machine Tools** (1,000,000 units) | Cars and Trucks (1,000,000 units) |
|---|---|---|---|---|---|---|---|---|---|
| 1978 | 0.533 | 0.380 | 0.679 | 1.238 | 9.36 | 0.2429 | 4.838 | 0.1832 | 0.1491 |
| 1979 | 0.537 | 0.435 | 0.793 | 1.407 | 11.69 | 0.2637 | 6.212 | 0.1396 | 0.1857 |
| 1980 | 0.537 | 0.490 | 0.898 | 1.520 | 11.46 | 0.1625 | 4.193 | 0.1336 | 0.2223 |
| 1981 | 0.484 | 0.505 | 0.916 | 1.513 | 7.29 | 0.1149 | 1.395 | 0.1026 | 0.1756 |
| 1982 | 0.457 | 0.565 | 1.003 | 1.675 | 8.64 | 0.1582 | 1.645 | 0.0998 | 0.1963 |
| 1983 | 0.331 | 0.654 | 1.121 | 1.808 | 12.71 | 0.2016 | 2.740 | 0.1210 | 0.2398 |
| 1984 | 0.299 | 0.648 | 1.180 | 1.846 | 15.69 | 0.2581 | 4.674 | 0.1335 | 0.3164 |
| 1985 | 0.211 | 0.652 | 1.234 | 1.953 | 19.26 | 0.3143 | 5.630 | 0.1672 | 0.4372 |
| 1986 | 0.203 | 0.695 | 1.322 | 2.151 | 19.17 | 0.2936 | 7.219 | 0.1637 | 0.3692 |
| 1987 | 0.260 | 0.839 | 1.571 | 2.345 | 24.48 | 0.2937 | 9.600 | 0.1463 | 0.4723 |

Note:
* Referring to those with generating capacity of 500 Kw and over, including water, steam and gas turbine generators.

## OUTPUT OF MAJOR INDUSTRIAL PRODUCTS
(continued)

| Year | Medium and Large Tractors* (1,000,000 units) | Walking Tractors (1,000,000 units) | Internal Combustion Engines** (Commodity) (1,000,000,000 watts) | Railway Locomotives (unit) | Railway Freight Cars (1,000,000 units) | Railway Passenger Cars (unit) | Steel Ships for Civilian Use (1,000,000 units) |
|---|---|---|---|---|---|---|---|
| 1978 | 0.1135 | 0.3242 | 20.73 | 521 | 0.0170 | 784 | 0.861 |
| 1979 | 0.1256 | 0.3175 | 21.39 | 573 | 0.0160 | 856 | 0.809 |
| 1980 | 0.0977 | 0.2179 | 18.67 | 512 | 0.0106 | 1002 | 0.818 |
| 1981 | 0.0528 | 0.1989 | 14.74 | 398 | 0.0088 | 1159 | 0.916 |
| 1982 | 0.0403 | 0.2983 | 16.89 | 486 | 0.0106 | 1153 | 1.025 |
| 1983 | 0.0370 | 0.4977 | 21.32 | 589 | 0.0158 | 1230 | 1.294 |
| 1984 | 0.0397 | 0.6886 | 29.95 | 658 | 0.0181 | 1200 | 1.654 |
| 1985 | 0.0450 | 0.8225 | 40.80 | 746 | 0.0193 | 1447 | 2.219 |
| 1986 | 0.0286 | 0.7726 | 44.06 | 818 | 0.0206 | 1522 | 1.757 |
| 1987 | 0.0398 | 1.0972 | 43.96 | 909 | 0.0216 | 1791 | 1.979 |

Note:
* Referring to both wheel and crawler tractors with a haulage capacity of 14.72 Kw and over.

** Refers to the part of products sold in markets rather than distributed by government planning.

## 4.5 TOTAL ENERGY PRODUCTION

| Year | Total Output (equivalent to 1,000,000 tons of SCE*) | Percentage of Total Output | | | |
|---|---|---|---|---|---|
| | | Raw Coal | Crude Oil | Natural Gas | Hydropower |
| 1978 | 627.70 | 70.3 | 23.7 | 2.9 | 3.1 |
| 1979 | 645.62 | 70.2 | 23.5 | 3.0 | 3.3 |
| 1980 | 637.21 | 69.4 | 23.8 | 3.0 | 3.8 |
| 1981 | 632.23 | 70.2 | 22.9 | 2.7 | 4.2 |
| 1982 | 667.72 | 71.2 | 21.9 | 2.4 | 4.5 |
| 1983 | 712.63 | 71.6 | 21.3 | 2.3 | 4.8 |
| 1984 | 778.47 | 72.4 | 21.1 | 2.1 | 4.4 |
| 1985 | 855.38 | 72.8 | 20.9 | 2.0 | 4.3 |
| 1986 | 881.35 | 72.4 | 21.2 | 2.1 | 4.3 |
| 1987 | 891.18 | 72.0 | 21.5 | 2.1 | 4.4 |

Note:
* SCE refers to standard coal equivalent. The conversion is as follows:
1 kg coal = 0.714 kg SCE
1 kg crude oil = 1.43 kg SCE
1 cubic meter natural gas = 1.33 kg SCE.

## 4.6 *OUTPUT OF MAJOR INDUSTRIAL PRODUCTS BY REGION 1987*

| Region | Steel (1,000,000 tn) | Pig Iron (1,000,000 tn) | Rolled Steel Final Products (1,000,000 tn) | Crude Coal (1,000,000 tn) | Crude Oil (1,000,000 tn) | Electricity (1,000,000,000 Kwh) | Cement (1,000,000 tn) | Sulphuric Acid (1,000,000 tn) | Soda Ash (1,000,000 tn) |
|---|---|---|---|---|---|---|---|---|---|
| ***National Total*** | **56.017** | **54.324** | **43.914** | **928.083** | **134.041** | **497.27** | **181.283** | **9.620** | **2.372** |
| *Beijing* | 3.354 | 3.294 | 2.831 | 8.998 | | 10.58 | 2.993 | 0.119 | |
| *Tianjin* | 1.634 | 1.015 | 1.750 | | 4.531 | 8.11 | 1.146 | 0.113 | 0.541 |
| *Hebei* | 2.859 | 3.679 | 2.443 | 63.426 | 7.954 | 31.26 | 11.393 | 0.460 | 0.023 |
| *Shanxi* | 2.007 | 3.093 | 1.187 | 230.937 | | 26.34 | 5.445 | 0.296 | 0.031 |
| *Inner Mongolia* | 2.155 | 2.569 | 1.139 | 34.102 | | 12.65 | 2.220 | 0.038 | 0.033 |
| *Liaoning* | 11.272 | 11.462 | 8.057 | 43.186 | 11.454 | 38.44 | 11.628 | 0.742 | 0.712 |
| *Jilin* | 0.392 | 0.486 | 0.534 | 21.087 | 2.865 | 10.63 | 3.731 | 0.114 | 0.021 |
| *Heilongjiang* | 0.800 | 0.388 | 0.700 | 68.208 | 55.553 | 22.35 | 4.944 | 0.124 | |
| *Shanghai* | 8.669 | 4.969 | 4.738 | | | 27.63 | 2.303 | 0.348 | 0.026 |
| *Jiangsu* | 1.300 | 1.351 | 1.720 | 22.351 | 0.625 | 30.12 | 15.110 | 0.896 | 0.052 |
| *Zhejiang* | 0.564 | 0.354 | 0.841 | 1.443 | | 17.31 | 11.896 | 0.283 | 0.076 |
| *Anhui* | 2.108 | 2.674 | 1.646 | 28.843 | | 16.88 | 7.563 | 0.476 | 0.010 |
| *Fujian* | 0.389 | 0.449 | 0.336 | 7.872 | | 9.85 | 3.748 | 0.178 | 0.004 |
| *Jiangxi* | 1.053 | 0.840 | 0.907 | 19.678 | | 10.28 | 4.282 | 0.359 | 0.002 |
| *Shandong* | 1.632 | 2.113 | 1.228 | 53.177 | 31.602 | 33.18 | 14.819 | 0.537 | 0.327 |
| *Henan* | 1.132 | 1.392 | 1.090 | 80.622 | 9.320 | 25.93 | 9.252 | 0.261 | 0.046 |
| *Hubei* | 5.735 | 5.003 | 5.156 | 8.491 | 1.010 | 26.86 | 9.208 | 0.701 | 0.190 |
| *Hunan* | 1.133 | 1.354 | 1.069 | 33.601 | | 16.14 | 9.617 | 0.664 | 0.041 |
| *Guangdong* | 0.886 | 0.772 | 1.047 | 8.736 | 0.453 | 23.02 | 14.799 | 0.820 | |
| *Guangxi* | 0.418 | 0.412 | 0.459 | 7.965 | 0.038 | 10.26 | 5.429 | 0.278 | 0.015 |
| *Sichuan* | 4.283 | 3.738 | 3.227 | 61.325 | 0.117 | 26.29 | 12.439 | 0.874 | 0.182 |
| *Guizhou* | 0.320 | 0.727 | 0.249 | 31.696 | | 8.49 | 2.585 | 0.056 | |
| *Yunnan* | 0.607 | 0.825 | 0.497 | 19.459 | | 9.43 | 3.843 | 0.324 | |
| *Tibet* | 0.006 | 0.026 | | 0.006 | | 0.26 | 0.046 | | |
| *Shaanxi* | 0.397 | 0.344 | 0.266 | 28.556 | 0.357 | 14.37 | 4.398 | 0.192 | 0.009 |
| *Gansu* | 0.448 | 0.733 | 0.223 | 12.824 | 1.528 | 13.36 | 3.059 | 0.279 | 0.031 |
| *Qinghai* | 0.242 | 0.188 | 2.743 | 2.743 | 0.575 | 2.24 | 0.482 | 0.020 | |
| *Ningxia* | 0.021 | 0.044 | 0.040 | 12.967 | 0.308 | 4.26 | 0.729 | 0.031 | |
| *Xinjiang* | 0.207 | 0.244 | 0.166 | 15.784 | 5.751 | 4.86 | 2.176 | 0.037 | |

## *OUTPUT OF MAJOR INDUSTRIAL PRODUCTS BY REGION 1987*
(continued)

| Region | Caustic Soda (1,000,000 tn) | Synthetic Ammonia (1,000,000 tn) | Chemical Fertilizers (1,000,000 tn) | Chemical Pesticides (1,000,000 tn) | Chemical Fibers (1,000,000 tn) | Yarn (1,000,000 tn) | Cloth (1,000,000,000 m) | Machine Made Paper, Paperboard (1,000,000 tn) | Sugar (1,000,000 tn) |
|---|---|---|---|---|---|---|---|---|---|
| ***National Total*** | **2.735** | **19.487** | **17.033** | **0.026** | **1.157** | **4.324** | **16.71** | **10.082** | **5.108** |
| *Beijing* | 0.096 | 0.108 | 0.109 | | 0.038 | 0.077 | 0.28 | 0.193 | 0.024 |
| *Tianjin* | 0.247 | 0.085 | 0.065 | 0.024 | 0.070 | 0.126 | 0.44 | 0.243 | |
| *Hebei* | 0.084 | 1.575 | 1.248 | 0.014 | 0.017 | 0.272 | 1.12 | 0.494 | 0.006 |
| *Shanxi* | 0.069 | 0.741 | 0.597 | 0.003 | 0.017 | 0.103 | 0.48 | 0.214 | 0.034 |
| *Inner Mongolia* | 0.031 | 0.191 | 0.126 | | 0.004\ | 0.019 | 0.09 | 0.106 | 0.172 |
| *Liaoning* | 0.251 | 0.862 | 0.657 | 0.013 | 0.141 | 0.233 | 0.67 | 0.783 | 0.028 |
| *Jilin* | 0.071 | 0.479 | 0.293 | 0.001 | 0.025 | 0.058 | 0.20 | 0.544 | 0.052 |
| *Heilongjiang* | 0.064 | 0.494 | 0.361 | 0.001 | 0.047 | 0.085 | 0.24 | 0.485 | 0.385 |
| *Shanghai* | 0.283 | 0.354 | 0.234 | 0.010 | 0.234 | 0.397 | 1.48 | 0.442 | |
| *Jiangsu* | 0.279 | 1.739 | 1.474 | 0.057 | 0.230 | 0.598 | 2.42 | 0.638 | 0.022 |
| *Zhejiang* | 0.137 | 0.922 | 0.752 | 0.025 | 0.057 | 0.193 | 0.54 | 0.594 | 0.008 |
| *Anhui* | 0.072 | 1.088 | 0.961 | 0.005 | 0.014 | 0.167 | 0.66 | 0.304 | |
| *Fujian* | 0.079 | 0.482 | 0.403 | 0.009 | 0.018 | 0.057 | 0.21 | 0.425 | 0.445 |
| *Jiangxi* | 0.053 | 0.257 | 0.304 | 0.003 | 0.017 | 0.086 | 0.34 | 0.246 | 0.117 |
| *Shandong* | 0.187 | 1.555 | 1.282 | 0.015 | 0.018 | 0.412 | 1.58 | 0.722 | 0.006 |
| *Henan* | 0.091 | 1.356 | 1.195 | 0.003 | 0046 | 0.281 | 1.07 | 0.542 | |
| *Hubei* | 0.080 | 1.1170 | 1.138 | 0.011 | 0.021 | 0.384 | 1.37 | 0.404 | 0.008 |
| *Hunan* | 0.127 | 1.273 | 1.165 | 0.026 | 0.028 | 0.125 | 0.48 | 0.544 | 0.062 |
| *Guangdong* | 0.094 | 0.708 | 0.730 | 0.018 | 0.057 | 0.100 | 0.29 | 0.667 | 1.856 |
| *Guangxi* | 0.044 | 0.360 | 0.376 | 0.004 | 0.008 | 0.058 | 0.19 | 0.215 | 1.033 |
| *Sichuan* | 0.169 | 1.627 | 1.544 | 0.008 | 0.024 | 0.107 | 0.81 | 0.702 | 0.161 |
| *Guizhou* | 0.014 | 0.385 | 0.412 | 0.001 | | 0.026 | 0.09 | 0.046 | 0.005 |
| *Yunnan* | 0.023 | 0.586 | 0.708 | 0.001 | 0.002 | 0.041 | 0.17 | 0.131 | 0.523 |
| *Tibet* | | | | | | | | | |
| *Shaanxi* | 0.041 | 0.413 | 0.405 | | 0.011 | 0.171 | 0.73 | 0.234 | 0.002 |
| *Gansu* | 0.024 | 0.329 | 0.210 | 0.008 | 0.011 | 0.017 | 0.06 | 0.070 | 0.044 |
| *Qinghai* | 0.007 | 0.006 | 0.035 | | | 0.005 | 0.02 | 0.006 | |
| *Ningxia* | 0.005 | 0.104 | 0.086 | | 0.002 | 0.005 | 0.02 | 0.038 | 0.028 |
| *Xinjiang* | 0.007 | 0.232 | 0.163 | | | 0.061 | 0.26 | 0.050 | 0.087 |

## 4.7 *TOTAL ENERGY CONSUMPTION*

| Year | Total Domestic Consumption (1,000,000 tn of SCE) | Percentage of Total Consumption | | | |
|---|---|---|---|---|---|
| | | Coal | Petroleum | Natural Gas | Hydro-electric power |
| *1979* | 585.88 | 71.31 | 21.79 | 3.30 | 3.60 |
| *1980* | 602.75 | 72.10 | 20.85 | 3.06 | 3.99 |
| *1981* | 594.47 | 72.75 | 20.00 | 2.74 | 4.51 |
| *1982* | 626.46 | 74.02 | 18.68 | 2.48 | 4.82 |
| *1983* | 660.40 | 74.29 | 18.05 | 2.40 | 5.26 |
| *1984* | 709.04 | 75.31 | 17.45 | 2.33 | 4.91 |
| *1985* | 770.20 | 75.92 | 17.02 | 2.23 | 4.83 |
| *1986* | 808.82 | 76.03 | 17.06 | 2.26 | 4.65 |
| *1987* | 858.15 | 76.14 | 17.03 | 2.15 | 4.68 |

## 4.8 *OVERALL LABOR PRODUCTIVITY OF INDUSTRIAL ENTERPRISES**

(at 1980 constant prices)

| Year | Labor Productivity (yuan/person-year) | Index (1978=100) |
|---|---|---|
| *1978* | 11131 | 100.0 |
| *1979* | 11838 | 106.4 |
| *1980* | 12081 | 108.5 |
| *1981* | 11863 | 106.6 |
| *1982* | 12133 | 109.0 |
| *1983* | 13049 | 117.2 |
| *1984* | 14070 | 126.4 |
| *1985* | 15080 | 135.5 |
| *1986* | 15809 | 142.0 |
| *1987* | 16889 | 151.7 |

**Note:**
*** Figures in this table only include state owned units with independent accounting system.**

## 4.9 *OVERALL LABOR PRODUCTIVITY OF SELECTED INDUSTRIES*

(yuan/person-year)

| Industries | 1986 |
|---|---|
| *Coal Mining and Preparation* | 3588 |
| *Petroleum and Natural Gas Extraction* | 26050 |
| *Ferrous Metals Mining and Preparation* | 6355 |
| *Non-ferrous Metals Mining and Preparation* | 6340 |
| *Building Materials and Other Non-metals Mineral Preparation* | 4138 |
| *Food Manufacturing* | 24121 |
| *Beverage Industry* | 14211 |
| *Tobacco Manufacturing* | 74763 |
| *Textile Manufacturing* | 18439 |
| *Sewing Industry* | 18502 |
| *Leather, Furs and Related Products* | 13915 |
| *Timber Processing and Bamboo, Rattan, Palm and Straw Products* | 9965 |
| *Paper Making and Manufactured Goods* | 15743 |
| *Power Generation, Steam and Hot Water Production and Supply* | 24728 |
| *Petroleum Processing* | 83556 |
| *Chemical Industry* | 18556 |
| *Medical and Pharmaceutical Products* | 28422 |
| *Chemical Fibers* | 40313 |
| *Rubber Manufacturing* | 29043 |
| *Plastics Manufacturing* | 24023 |
| *Building Material and Other Non-Metals Mineral Products* | 7343 |
| *Smelting and Pressing of Ferrous Metals* | 18424 |
| *Smelting and Pressing of Non-Ferrous* | 26468 |
| *Metals Products* | 13931 |
| *Engineering Industry* | 11262 |
| *Transportation Equipment Manufacturing* | 12212 |
| *Electric Equipment and Machinery* | 19762 |
| *Electric and Telecommunication Equipment* | 23022 |
| *Precision Instrument Manufacturing* | 10371 |

**Note:**
*** Figures in this table only include state owned units with independent accounting.**

*Section V*

## TRANSPORTATION, POSTAL AND TELECOMMUNICATION SERVICES

*TRANSPORTATION, POSTAL AND TELECOMMUNICATION SERVICES*

## 5.1 *LENGTH OF VARIOUS TRANSPORTATION ROUTES*
(1,000,000 km)

| Year | Operating Railways* | Highways | Inland Navigable Waterways | Civil Aviation Routes | | Petroleum and Gas Pipelines |
|---|---|---|---|---|---|---|
| | | | | Total | International Routes | |
| *1978* | 0.0486 | 0.8902 | 0.1360 | 0.1489 | 0.0553 | 0.0083 |
| *1979* | 0.0498 | 0.8758** | 0.1078** | 0.1600 | 0.0513 | 0.0091 |
| *1980* | 0.0499 | 0.8883 | 0.1085 | 0.1917 | 0.0812 | 0.0087 |
| *1981* | 0.0502 | 0.8975 | 0.1087 | 0.2182 | 0.0828 | 0.0097 |
| *1982* | 0.0505 | 0.9070 | 0.1086 | 0.2327 | 0.0999 | 0.0104 |
| *1983* | 0.0516 | 0.9151 | 0.1089 | 0.2291 | 0.0999 | 0.0109 |
| *1984* | 0.0517 | 0.9267 | 0.1093 | 0.2602 | 0.1074 | 0.0111 |
| *1985* | 0.0521 | 0.9424 | 0.1091 | 0.2772 | 0.1060 | 0.0118 |
| *1986* | 0.0525 | 0.9628 | 0.1094 | 0.3243 | 0.1076 | 0.0130 |
| *1987* | 0.0526 | 0.9813 | 0.1094 | 0.3871 | 0.1489 | 0.0134 |

**Note:**
*** Figures for operating railways exclude local railways.**

**** Refer to census statistics at the end of October.**

## 5.2 *VOLUME OF FREIGHT TRAFFIC MEASURED IN TONS**
(1,000,000 tons)

| Year | Total Traffic | Railways | Highways | | Waterways | Petroleum and Gas Pipelines | Civil Aviation |
|---|---|---|---|---|---|---|---|
| | | | Total | Transportation Department | | | |
| *1978* | 2489.46 | 1101.19 | 851.82 | 851.82 | 432.92 | 103.47 | 0.064 |
| *1979* | 5375.08 | 1118.93 | 3710.36 | 815.56 | 432.29 | 113.42 | 0.080 |
| *1980* | 5465.37 | 1112.79 | 3820.48 | 760.17 | 426.76 | 105.25 | 0.089 |
| *1981* | 5237.64 | 1076.73 | 3636.63 | 715.04 | 414.90 | 109.29 | 0.094 |
| *1982* | 5479.35 | 1135.32 | 3792.05 | 787.77 | 443.29 | 108.59 | 0.102 |
| *1983* | 5768.87 | 1187.84 | 4014.13 | 790.78 | 450.58 | 116.20 | 0.116 |
| *1984* | 7169.07 | 1240.74 | 5333.82 | 788.68 | 468.92 | 125.44 | 0.150 |
| *1985* | 7457.62 | 1307.08 | 5380.62 | 762.27 | 633.22 | 136.50 | 0.195 |
| *1986* | 8535.58 | 1356.35 | 6201.13 | 785.49 | 829.62 | 148.25 | 0.224 |
| *1987* | 9095.66 | 1406.53 | 6610.14 | 711.72 | 937.55 | 141.14 | 0.299 |

**Note:**
*** Figures of freight traffic by highway after 1979 and by waterway after 1985 include all state owned, collective owned and individual owned units. Data before these periods only include state owned units.**

## 5.3 *VOLUME OF FREIGHT TRAFFIC MEASURED IN TONS-KM**

(1,000,000,000 tn-km)

| Year | Total Traffic | Railways | Highways | | Waterways | Petroleum and Gas Pipelines | Civil Aviation |
|---|---|---|---|---|---|---|---|
| | | | Total | Transportation Department | | | |
| *1978* | 982.9 | 534.5 | 27.4 | 27.4 | 377.9 | 43.0 | 0.10 |
| *1979* | 1138.4 | 559.8 | 74.5 | 26.8 | 456.4 | 47.6 | 0.12 |
| *1980* | 1202.6 | 571.7 | 76.4 | 25.5 | 505.3 | 49.1 | 0.14 |
| *1981* | 1214.3 | 571.2 | 78.0 | 25.3 | 515.0 | 49.9 | 0.17 |
| *1982* | 1304.9 | 612.0 | 94.9 | 30.3 | 547.7 | 50.1 | 0.20 |
| *1983* | 1404.4 | 664.6 | 108.4 | 33.5 | 578.8 | 52.4 | 0.23 |
| *1984* | 1569.4 | 724.8 | 153.6 | 35.4 | 633.5 | 57.2 | 0.31 |
| *1985* | 1812.6 | 812.6 | 169.3 | 35.4 | 770.0 | 60.3 | 0.42 |
| *1986* | 2014.8 | 876.5 | 211.8 | 36.9 | 864.8 | 61.2 | 0.48 |
| *1987* | 2192.7 | 947.2 | 240.7 | 38.8 | 941.6 | 62.5 | 0.71 |

**Note:**
*** Figures of volume of freight traffic by highway after 1979 and by waterway after 1985 include all state owned, collectively owned and privately owned units. The data before these periods only include state owned units.**

## 5.4 *DISTRIBUTION OF FREIGHT TRAFFIC VOLUME**

(total = 100)

| Year | Volume of Freight Traffic in Tons | | | Volume of Freight Traffic in Tons-km | | |
|---|---|---|---|---|---|---|
| | Railways | Highways | Waterways | Railways | Highways | Waterways |
| *1978* | 44.2 | 34.2 | 17.4 | 54.4 | 2.8 | 38.5 |
| *1979* | 20.8 | 69.0 | 8.0 | 49.2 | 6.5 | 40.1 |
| *1980* | 20.4 | 69.9 | 7.8 | 47.5 | 6.4 | 42.0 |
| *1981* | 20.6 | 69.4 | 7.9 | 47.0 | 6.4 | 42.4 |
| *1982* | 20.7 | 69.2 | 8.1 | 46.9 | 7.3 | 42.0 |
| *1983* | 20.6 | 69.6 | 7.8 | 47.3 | 7.7 | 41.2 |
| *1984* | 17.3 | 74.4 | 6.5 | 46.2 | 9.8 | 40.4 |
| *1985* | 17.5 | 72.1 | 8.5 | 44.8 | 9.3 | 42.5 |
| *1986* | 15.9 | 72.7 | 9.7 | 43.5 | 10.5 | 42.9 |
| *1987* | 15.5 | 72.7 | 10.3 | 43.2 | 11.0 | 42.9 |

**Note:**
*** Figures for freight traffic volume by highway after 1979 and by waterway after 1985 include all state owned, collective owned and private owned units. The data before these periods only includes state owned units.**

## 5.5 *VOLUME OF PASSENGER TRAFFIC MEASURED IN NUMBER OF PERSONS**

(1,000,000 persons)

| Year | Total Traffic | Railways | Highways | Waterways | Civil Aviation |
|---|---|---|---|---|---|
| *1978* | 2539.93 | 814.91 | 1492.29 | 230.42 | 2.31 |
| *1979* | 2896.66 | 863.90 | 1786.18 | 243.60 | 2.98 |
| *1980* | 3417.85 | 922.04 | 2227.99 | 264.39 | 3.43 |
| *1981* | 3848.44 | 953.00 | 2615.59 | 275.84 | 4.01 |
| *1982* | 4289.63 | 999.21 | 3006.10 | 279.87 | 4.45 |
| *1983* | 4706.14 | 1060.44 | 3369.65 | 272.14 | 3.91 |
| *1984* | 5302.17 | 1133.53 | 3903.36 | 259.74 | 5.54 |
| *1985* | 6202.06 | 1121.10 | 4764.86 | 308.63 | 7.47 |
| *1986* | 6882.12 | 1085.79 | 5442.59 | 343.77 | 9.96 |
| *1987* | 7336.40 | 1124.79 | 5855.82 | 342.69 | 13.10 |

**Note:**
*** Figures for freight traffic volume by highway after 1979 and by waterway after 1985 include all state owned, collective owned and private owned units. The data before these periods only includes state owned units.**

## 5.6 *VOLUME OF PASSENGER TRAFFIC MEASURED IN NUMBER OF PERSONS-KM**

(1,000,000,000 persons-km)

| Year | Total Traffic | Railways | Highways | Waterways | Civil Aviation |
|---|---|---|---|---|---|
| *1978* | 174.3 | 109.3 | 52.1 | 10.1 | 2.8 |
| *1979* | 196.8 | 121.6 | 60.3 | 11.4 | 3.5 |
| *1980* | 228.1 | 138.3 | 72.9 | 12.9 | 4.0 |
| *1981* | 250.0 | 147.3 | 83.9 | 13.8 | 5.0 |
| *1982* | 274.4 | 157.5 | 96.4 | 14.5 | 6.0 |
| *1983* | 309.5 | 177.6 | 110.6 | 15.4 | 5.9 |
| *1984* | 362.0 | 204.6 | 133.7 | 15.4 | 8.3 |
| *1985* | 443.7 | 241.6 | 172.5 | 17.9 | 11.7 |
| *1986* | 489.7 | 258.7 | 198.2 | 18.2 | 14.6 |
| *1987* | 542.7 | 284.3 | 220.4 | 19.3 | 18.7 |

**Note:**
*** Figures for freight traffic volume by highway after 1979 and by waterway after 1985 include all state owned, collective owned and private owned units. The data before these periods only includes state owned units.**

## 5.7 *CARGO HANDLED AT MAJOR SEAPORTS*
(1,000,000 tn)

| Port | 1978 | 1980 | 1985 | 1986 | 1987 |
|---|---|---|---|---|---|
| ***Total*** | ***198.34** | ***217.31** | ***311.54** | **371.80** | **396.81** |
| *Dalian* | 286.64 | 32.63 | 43.81 | 44.29 | 46.08 |
| *Yingkou* | 0.33 | 0.24 | 0.98 | 1.13 | 1.15 |
| *Qinhuangdao* | 22.19 | 26.41 | 44.19 | 48.73 | 53.78 |
| *Tianjin* | 11.31 | 11.92 | 18.56 | 18.18 | 17.23 |
| *Yantai* | 4.58 | 5.06 | 6.89 | 6.91 | 6.94 |
| *Qingdao* | 20.02 | 17.08 | 26.11 | 28.01 | 30.28 |
| *Shijiugang* | | | | 2.63 | 4.25 |
| *Lianyungang* | 5.94 | 7.39 | 9.29 | 9.49 | 8.94 |
| *Shanghai* | 79.55 | 84.83 | 112.91 | 126.04 | 128.08 |
| *Ningbo* | | 3.26 | 10.40 | 17.95 | 19.43 |
| *Shantou* | 1.53 | 1.76 | 2.01 | 2.31 | 2.50 |
| *Huangpu* | 10.50 | 12.10 | 17.72 | 19.17 | 26.05 |
| *Zhanjiang* | 9.47 | 10.75 | 12.31 | 12.96 | 14.14 |
| *Haikou* | 0.76 | 0.72 | 1.70 | 1.74 | |
| *Basuo* | 3.07 | 2.78 | 3.88 | 3.91 | 6.67 |
| *Sanya* | 0.45 | 0.38 | 0.78 | 0.64 | |
| *Other Middle Sized Ports* | | | | 27.71 | 31.11 |

**Note:**
***The total of 1978-1985 does not include other middle sized ports.**

## 5.8 *PRINCIPAL ECONOMIC AND TECHNICAL INDICATORS OF RAILWAY, HIGHWAY AND WATERWAY TRAFFIC*

| Item | | 1986 | 1987 |
|---|---|---|---|
| ***Railways*** | | | |
| *Turnover Period of Freight Wagons* | day | 3.57 | 3.77 |
| *Carrying Capacity of Freight Wagons* | ton | 52.2 | 53.2 |
| *Average Traveling Speed of Freight Wagons* | km/hour | 28.5 | 28.3 |
| *Average Daily Haul of Freight Locomotives* | 1,000,000 tons/km | 0.806 | 0.821 |
| *Average Daily Haul of Diesel Locomotives* | 1,000,000 tons/km | 0.906 | 0.961 |
| *Coal Consumption of Steam Locomotives* | kg/1,000,000 tons-km | 10820 | 11090 |
| *Oil Consumption of Diesel Locomotives* | kg/1,000,000 tons-km | 2690 | 2590 |
| ***Highways*** | | | |
| *Percentage of Trucks in Good Condition* | % | 86.7 | 86.6 |
| *Annual Cargo Carried per Ton of Truck Capacity** | tons-km | 34178 | 34017 |
| *Annual Capacity per Truck* | tons-km | 198719 | 204844 |
| *Gasoline Consumption per Truck* | liter/1000 tons-km | 76 | 76 |
| *Diesel Oil Consumption per Truck* | liter/1000 tons-km | 51 | 51 |
| ***Waterways*** | | | |
| *Operating Rate of Cargo Vessels** under the Ministry of Transportation* | % | 91.6 | 90.7 |
| *Annual Cargo Carried per Ton of Freighter Capacity* under the Ministry of Transportation* | tons-km | 46676 | 50528 |
| *Annual cargo Carried per Kw of Tugboat Capacity under the Ministry of Transportation* | ton-km | 70650 | 70059 |
| *Annual Cargo Carried per Ton of Barge Capacity under the Ministry of Transportation* | tons | 15523 | 15001 |
| *Operating Rate of Local Inland Water Tugboats*** | % | 85.5 | 86.7 |
| *Annual Cargo Carried by Local Inland WaterTugboats* | ton-km/Kw capacity | 56208 | 57505 |

**Note:**
*** Cargo carried per ton of truck or freighter capacity equals tons-km of cargo carried divided by total ton capacity of all trucks or freighters.**

**** Operating rate of cargo vessels equals cargo carried multiplied by working days divided by the capacity tons of all vessels multiplied by number of calendar days.**

## 5.9 NUMBER OF POSTAL AND TELECOMMUNICATION OFFICES AND LENGTH OF POSTAL ROUTES

| Year | Post Offices (1,000,000) | | Length of Postal Routes (1,000,000 km) | | Long Distance Telephone Lines | Telegraph Lines |
|---|---|---|---|---|---|---|
| | Total | Rural Post Offices | Total | Air Routes | | |
| 1978 | 0.0496 | 0.0431 | 4.8633 | 0.1465 | 18801 | 8430 |
| 1979 | 0.0496 | 0.0430 | 4.8123 | 0.1616 | 20307 | 8785 |
| 1980 | 0.0495 | 0.0428 | 4.7371 | 0.1633 | 22011 | 9146 |
| 1981 | 0.0496 | 0.0426 | 4.6602 | 0.1585 | 23909 | 8608 |
| 1982 | 0.0497 | 0.0426 | 4.6763 | 0.1619 | 25961 | 9178 |
| 1983 | 0.0502 | 0.0427 | 4.7209 | 0.1515 | 28637 | 9467 |
| 1984 | 0.0515 | 0.0438 | 4.8306 | 0.1778 | 32354 | 9685 |
| 1985 | 0.0531 | 0.0450 | 4.9821 | 0.2265 | 37551 | 9916 |
| 1986 | 0.0528 | 0.0445 | 5.0420 | 0.2883 | 44085 | 10149 |
| 1987 | 0.0529 | 0.0443 | 5.0330 | 0.2923 | 53416 | 10564 |

## 5.10 VOLUME OF POSTAL AND TELECOMMUNICATION SERVICES BY TYPE OF SERVICE

| Year | Postal and Telecommunication Services (1,000,000,000 yuan) | Number of Letters Through Post Offices (1,000,000,000) | Number of Newspapers, Magazines Circulated per Issue (1,000,000) | Number of Telegrams (1,000,000) | Long Distance Telephone Calls (1,000,000) | Number of Urban Telephone (1,000,000) |
|---|---|---|---|---|---|---|
| 1978 | 1.165 | 2.835 | 112.50 | 127.48 | 185.74 | 1.1915 |
| 1979 | 1.255 | 3.080 | 126.80 | 134.95 | 205.87 | 1.2702 |
| 1980 | 1.334 | 3.313 | 164.31 | 146.63 | 214.04 | 1.3417 |
| 1981 | 1.402 | 3.388 | 181.24 | 159.38 | 220.49 | 1.4264 |
| 1982 | 2.041 | 3.394 | 195.98 | 160.71 | 235.74 | 1.5387 |
| 1983 | 2.226 | 3.521 | 229.33 | 180.75 | 265.56 | 1.6886 |
| 1984 | 2.503 | 3.948 | 281.41 | 174.14 | 315.53 | 1.9107 |
| 1985 | 2.960 | 4.678 | 301.72 | 218.98 | 382.54 | 2.1896 |
| 1986 | 3.286 | 4.959 | 287.31 | 211.09 | 423.03 | 2.5051 |
| 1987 | 3.884 | 5.479 | 310.05 | 250.27 | 515.25 | 2.9304 |

**Note:**
**The postal and telecommunication services in 1982-1986 are calculated at 1980 constant prices and those of the preceding years are calculated at 1970 constant prices.**

## 5.11 *VOLUME OF POSTAL AND TELECOMMUNICATION SERVICES BY YEAR*

| Item | 1980 | 1985 | 1986 | 1987 |
|---|---|---|---|---|
| ***National Levels*** | | | | |
| *Annual Average Number of Letters Mailed per Capita* | 3.4 | 4.5 | 4.7 | 5.1 |
| *Annual Average Number of Newspaper and Magazine Subscribers per 100 Persons* | 16.7 | 28.7 | 27.0 | 28.7 |
| *Percentage of Provincial and Municipal Districts that Receive China Daily on Day of Publication* | 69.0 | 72.4 | 75.9 | 75.9 |
| *Average Number of Telephones per 100 persons* | 0.4 | 0.6 | 0.7 | 0.8 |
| *Average Number of Long Distance Telephone Lines from Beijing to Provincial Capitals* | 27.1 | 43.7 | 67.2 | 73.9 |
| *Percentage of Cities at County Level and Above Equipped with Automatic Telephone Exchanges* | 35.7 | 40.3 | 44.1 | 49.4 |

## 5.12 *EXTENT OF POSTAL AND TELECOMMUNICATION SERVICES IN RURAL AREAS*

| Item | 1980 | 1985 | 1986 | 1987 |
|---|---|---|---|---|
| ***Percentage Level of Postal and Telecommunication Services In Rural Areas*** | | | | |
| *Percentage of Townships with Postal and Telecommunication Offices* | 66.4 | 64.8 | 64.0 | 66.8 |
| *Percentage of Townships with Postal Routes* | 99.6 | 99.6 | 99.7 | 99.3 |
| *Percentage of Villages with Postal Routes* | 96.5 | 96.3 | 96.1 | 95.8 |
| *Percentage of Municipalities and Counties that Receive the Provincial Newspaper on Day of Publication* | 42.6 | 39.9 | 39.0 | 34.1 |
| *Percentage of Townships that Receive the Provincial Newspaper on Day of Publication* | 19.1 | 16.7 | 16.4 | 14.7 |
| *Percentage of Villages that Receive the Provincial Newspaper on Day of Publication* | 9.5 | 10.1 | 9.7 | 8.8 |
| *Percentage of Townships Equipped with Telephone Switchboards* | 74.6 | 69.6 | 65.5 | 72.3 |
| *Percentage of Townships Equipped with Telephones* | 95.5 | 94.8 | 95.5 | 95.1 |
| *Percentage of Villages Equipped with Telephones* | 62.8 | 45.4 | 45.4 | 44.8 |

*Section VI*

# INVESTMENT IN FIXED ASSETS

## 6.1 *TOTAL INVESTMENT IN FIXED ASSETS OF SOCIETY*

(1,000,000,000 yuan)

| Item | 1981 | 1985 | 1986 | 1987 |
|---|---|---|---|---|
| **Total** | **96.101** | **254.319** | **301.962** | **351.811** |
| ***State Owned Units*** | 66.751 | 168.051 | 197.850 | 226.195 |
| *Capital Construction* | 44.291 | 107.437 | 117.611 | 132.410 |
| *Technical Updating and Transformation* | 19.530 | 44.914 | 61.921 | 74.263 |
| *Other* | 2.930 | 15.700 | 18.318 | 19.522 |
| ***Collective Owned Units*** | 11.524 | 32.746 | 39.174 | 48.066 |
| *Cities and Towns* | 3.157 | 12.823 | 14.639 | 18.232 |
| *Rural* | 8.367 | 19.923 | 24.535 | 29.834 |
| ***Individual Investments*** | 17.826 | 53.522 | 64.938 | 77.550 |
| *Urban* | 1.192 | 5.679 | 7.456 | 7.850 |
| *Rural* | 16.634 | 47.843 | 57.482 | 69.700 |

## 6.2 *TOTAL INVESTMENT IN FIXED ASSETS BY STATE OWNED UNITS*

(1,000,000,000 yuan)

| Year | Investment in Fixed Assets | | Investment in Capital Construction as Percentage of Investment in Fixed Assets |
|---|---|---|---|
| | Total | Investment in Capital Construction | |
| ***1976-1980*** | **318.622** | **234.217** | **73.5** |
| *1978* | 66.872 | 50.099 | 74.9 |
| *1979* | 69.936 | 52.348 | 74.9 |
| *1980* | 74.590 | 55.889 | 74.9 |
| ***1981-1985*** | **533.047** | **341.009** | **64.0** |
| *1981* | 66.751 | 44.291 | 66.4 |
| *1982* | 84.531 | 55.553 | 65.7 |
| *1983* | 95.196 | 59.413 | 62.4 |
| *1984* | 118.518 | 74.315 | 62.7 |
| *1985* | 168.051 | 107.437 | 63.9 |
| ***1986-1990*** | | | |
| *1986* | 197.850 | 117.611 | 59.4 |
| *1987* | 226.195 | 132.410 | 58.5 |

## 6.3 *TOTAL INVESTMENT IN FIXED ASSETS OF SOCIETY BY REGION 1987*

(1,000,000,000 yuan)

| Region | Total Investment in Fixed Assets | State Owned Units | | | Collective Owned Units and Individual Investment in Urban and Rural Areas |
|---|---|---|---|---|---|
| | | *Subtotal* | *Capital Construction* | *Technical Updating and Transformation* | |
| ***National Total*** | **351.811** | **226.195** | **132.410** | **74.263** | **125.616** |
| *Beijing* | 13.572 | 11.613 | 7.164 | 4.299 | 1.959 |
| *Tianjin* | 7.705 | 6.172 | 2.946 | 2.776 | 1.533 |
| *Hebei* | 17.931 | 8.726 | 5.144 | 3.108 | 9.205 |
| *Shanxi* | 10.064 | 7.512 | 4.652 | 2.081 | 2.552 |
| *Inner Mongolia* | 5.132 | 3.718 | 2.223 | 1.219 | 1.414 |
| *Liaoning* | 22.248 | 17.242 | 8.607 | 7.558 | 5.006 |
| *Jilin* | 7.780 | 5.546 | 2.713 | 2.061 | 2.234 |
| *Heilongjiang* | 13.318 | 11.285 | 6.233 | 2.751 | 2.033 |
| *Shanghai* | 18.054 | 15.378 | 9.232 | 5.737 | 2.676 |
| *Jiangsu* | 30.994 | 12.424 | 7.537 | 4.211 | 18.570 |
| *Zhejiang* | 19.078 | 5.872 | 3.425 | 2.153 | 13.206 |
| *Anhui* | 7.281 | 5.534 | 3.228 | 2.063 | 1.747 |
| *Fujian* | 7.649 | 4.876 | 3.500 | 1.255 | 2.773 |
| *Jiangxi* | 5.800 | 3.226 | 1.831 | 1.275 | 2.574 |
| *Shandong* | 29.120 | 15.591 | 7.878 | 3.733 | 13.529 |
| *Henan* | 16.009 | 8.264 | 3.790 | 2.851 | 7.745 |
| *Hubei* | 12.921 | 7.702 | 4.636 | 2.675 | 5.219 |
| *Hunan* | 11.666 | 5.851 | 2.895 | 2.456 | 5.815 |
| *Guangdong* | 25.303 | 15.596 | 10.518 | 4.821 | 9.707 |
| *Guangxi* | 5.784 | 3.739 | 1.787 | 1.648 | 2.045 |
| *Sichuan* | 19.484 | 12.806 | 6.497 | 4.870 | 6.678 |
| *Guizhou* | 3.976 | 2.543 | 1.424 | 0.950 | 1.433 |
| *Yunnan* | 5.282 | 3.572 | 1.939 | 1.425 | 1.710 |
| *Tibet* | 0.476 | 0.417 | 0.401 | 0.016 | 0.059 |
| *Shaanxi* | 7.000 | 5.244 | 3.166 | 1.933 | 1.756 |
| *Gansu* | 4.307 | 3.533 | 2.139 | 1.305 | 0.774 |
| *Qinghai* | 2.073 | 1.897 | 1.492 | 0.191 | 0.176 |
| *Ningxia* | 1.834 | 1.463 | 1.044 | 0.388 | 0.371 |
| *Xinjiang* | 5.613 | 4.495 | 2.620 | 0.904 | 1.118 |
| *Non-Region Specific* | 14.361 | 14.361 | 11.750 | 1.553 | |

## 6.4 *NEWLY INCREASED FIXED ASSETS AND NUMBER OF LARGE AND MEDIUM SIZE PROJECTS PUT INTO USE**

| Year | Newly Added Fixed Assets (1,000,000,000 yuan) | Percentage Rate of Fixed Assets Put into Use** | Number of Large and Medium Size Projects Put into Use | Percentage Rate of Large and Medium Sized Projects Put into Use |
|---|---|---|---|---|
| ***1976-1980**** | **174.731** | **74.6** | **515** | **7.4** |
| *1978* | 37.230 | 74.3 | 99 | 5.7 |
| *1979* | 43.802 | 83.7 | 128 | 9.7 |
| *1980* | 44.206 | 79.1 | 82 | 8.3 |
| ***1981-1985**** | **251.604** | **73.8** | **520** | **12.5** |
| *1981* | 38.340 | 86.6 | 79 | 10.6 |
| *1982* | 41.310 | 74.4 | 116 | 14.2 |
| *1983* | 45.310 | 76.3 | 91 | 11.2 |
| *1984* | 53.328 | 71.8 | 113 | 13.8 |
| *1985* | 73.316 | 68.2 | 121 | 12.6 |
| ***1986-1990**** | | | | |
| *1986* | 92.988 | 79.1 | 102 | 11.0 |
| *1987* | 88.642 | 66.9 | 102 | 14.6 |

**Note:**
*** The table refers to capital construction by state owned units.**

**** The percentage rate of fixed assets put into use is the ratio of newly increased fixed assets to the total investment in fixed assets. Statistics for the percentage rate of fixed assets put into use in 1966-1970 are incomplete, so the figure given is underestimated.**

***** The figure for the percentage rate of large and medium sized projects put into production in this line is the average of all years in that period.**

## 6.5 *FLOOR SPACE OF BUILDINGS COMPLETED**

| Year | Floor Space of Buildings Completed (1,000,000 sq. m.) | | Floor Space of Buildings Completed as Percentage of Total** | |
|---|---|---|---|---|
| | Total | Residential Buildings | Total | Residential Buildings |
| ***1976-1980*** | **500.40** | **234.86** | **49.8** | **52.4** |
| *1978* | 90.11 | 37.52 | 48.7 | 51.3 |
| *1979* | 120.00 | 62.56 | 50.4 | 52.1 |
| *1980* | 145.00 | 82.30 | 52.7 | 55.5 |
| ***1981-1985*** | **714.80** | **423.17** | **49.7** | **53.8** |
| *1981* | 129.41 | 79.04 | 51.4 | 53.5 |
| *1982* | 143.57 | 90.20 | 50.5 | 53.8 |
| *1983* | 132.12 | 81.25 | 52.3 | 58.0 |
| *1984* | 138.09 | 77.03 | 49.4 | 54.3 |
| *1985* | 171.61 | 95.65 | 46.3 | 50.6 |
| ***1986-1990*** | | | | |
| *1986* | 165.50 | 89.15 | 50.3 | 55.9 |
| *1987* | 142.11 | 64.97 | 47.9 | 53.9 |

**Note:**
*** The table refers to capital construction by state owned units.**

**** The total includes completed floor space of factory and storage buildings, offices, residential buildings, schools and hospitals.**

## 6.6 *TOTAL CAPITAL CONSTRUCTION INVESTMENT BY SECTOR**

(1,000,000,000 yuan)

| Year | Total | Industry | Construction Industry | Geological Prospecting | Agriculture, Forestry, Water Conservancy and Meteorology |
|---|---|---|---|---|---|
| ***1976-1980*** | **234.217** | **123.171** | **4.334** | **2.953** | **24.608** |
| *1978* | 50.099 | 27.316 | 0.884 | 1.165 | 5.334 |
| *1979* | 52.348 | 25.685 | 1.147 | 0.740 | 5.792 |
| *1980* | 55.889 | 27.561 | 1.122 | 0.303 | 5.203 |
| ***1981-1985*** | **341.009** | **154.697** | **6.317** | **1.942** | **17.284** |
| *1981* | 44.291 | 21.601 | 0.909 | 0.250 | 2.921 |
| *1982* | 55.553 | 26.060 | 1.052 | 0.259 | 3.412 |
| *1983* | 59.413 | 28.228 | 1.032 | 0.336 | 3.545 |
| *1984* | 74.315 | 34.159 | 1.124 | 0.372 | 3.712 |
| *1985* | 107.437 | 44.649 | 2.200 | 0.725 | 3.694 |
| ***1986-1990*** | | | | | |
| *1986* | 117.611 | 53.164 | 1.853 | 0.711 | 3.666 |
| *1987* | 132.410 | 66.602 | 1.854 | 0.794 | 4.370 |

Note:
* The table refers to capital construction by state owned units.

## *TOTAL CAPITAL CONSTRUCTION INVESTMENT BY SECTOR**

(continued)

| Year | Transportation, Postal, and Telecommunication Services | Commerce, Catering and Service Trades and Material Supplies | Scientific Research, Culture, Education, Public Health, and Social Welfare | Urban Public Utilities | Other |
|---|---|---|---|---|---|
| ***1976-1980*** | **30.245** | **8.735** | **12.783** | **9.512** | **17.876** |
| *1978* | 6.804 | 1.529 | 2.177 | 1.539 | 3.350 |
| *1979* | 6.409 | 2.057 | 3.347 | 2.991 | 4.180 |
| *1980* | 6.234 | 2.857 | 4.429 | 3.381 | 4.799 |
| ***1981-1985*** | **45.513** | **20.017** | **35.348** | **26.218** | **33.673** |
| *1981* | 4.047 | 2.801 | 4.363 | 3.185 | 4.214 |
| *1982* | 5.721 | 3.597 | 5.081 | 4.221 | 6.150 |
| *1983* | 7.804 | 2.893 | 5.944 | 3.802 | 5.829 |
| *1984* | 10.846 | 3.517 | 7.879 | 5.736 | 6.970 |
| *1985* | 17.095 | 7.209 | 12.081 | 9.274 | 10.510 |
| ***1986-1990*** | | | | | |
| *1986* | 18.081 | 4.192 | 14.028 | 11.000 | 10.916 |
| *1987* | 19.597 | 5.296 | 15.095 | 8.209 | 10.593 |

Note:
* The table refers to capital construction by state owned units.

## 6.7 DISTRIBUTION OF CAPITAL CONSTRUCTION BY SECTOR

| Year | Total | Industry | Construction Industry | Geological Prospecting | Agriculture, Forestry, Water Conservancy and Meteorology |
|---|---|---|---|---|---|
| ***1976-1980*** | **100.0** | **52.6** | **1.8** | **1.3** | **10.5** |
| *1978* | 100.0 | 54.5 | 1.8 | 2.3 | 10.6 |
| *1979* | 100.0 | 49.1 | 2.2 | 1.4 | 11.1 |
| *1980* | 100.0 | 49.3 | 2.0 | 0.5 | 9.3 |
| ***1981-1985*** | **100.0** | **45.4** | **1.8** | **0.6** | **5.0** |
| *1981* | 100.0 | 48.8 | 2.1 | 0.6 | 6.6 |
| *1982* | 100.0 | 46.9 | 1.9 | 0.5 | 6.1 |
| *1983* | 100.0 | 47.5 | 1.7 | 0.6 | 6.0 |
| *1984* | 100.0 | 46.0 | 1.5 | 0.5 | 5.0 |
| *1985* | 100.0 | 41.6 | 2.1 | 0.7 | 3.4 |
| ***1986-1990*** | | | | | |
| *1986* | 100.0 | 45.2 | 1.6 | 0.6 | 3.1 |
| *1987* | 100.0 | 50.3 | 1.4 | 0.6 | 3.3 |

### DISTRIBUTION OF CAPITAL CONSTRUCTION BY SECTOR
(continued)

| Year | Transportation, Postal, and Telecommunication Services | Commerce, Catering and Service Trades and Material Supplies | Scientific Research, Culture, Education, Public Health, and Social Welfare | Urban Public Utilities | Other |
|---|---|---|---|---|---|
| ***1976-1980*** | **12.9** | **3.7** | **5.5** | **4.1** | **7.6** |
| *1978* | 13.6 | 3.1 | 4.3 | 3.1 | 6.7 |
| *1979* | 12.2 | 3.9 | 6.4 | 5.7 | 8.0 |
| *1980* | 11.2 | 5.1 | 7.9 | 6.1 | 8.6 |
| ***1981-1985*** | **13.3** | **5.9** | **10.4** | **7.7** | **9.9** |
| *1981* | 9.1 | 6.3 | 9.8 | 7.2 | 9.5 |
| *1982* | 10.3 | 6.5 | 9.1 | 7.6 | 11.1 |
| *1983* | 13.1 | 4.9 | 10.0 | 6.4 | 9.8 |
| *1984* | 14.6 | 4.7 | 10.6 | 7.7 | 9.4 |
| *1985* | 15.9 | 6.7 | 11.2 | 8.6 | 9.8 |
| ***1986-1990*** | | | | | |
| *1986* | 15.4 | 3.6 | 11.9 | 9.3 | 9.3 |
| *1987* | 14.8 | 4.0 | 11.4 | 6.2 | 8.0 |

## 6.8 CAPITAL CONSTRUCTION INVESTMENT IN NEW PROJECTS, EXPANSION AND REBUILDING*

| Year | Investment (1,000,000,000 yuan) | | Percentage to Total Investment | |
|---|---|---|---|---|
| | New Projects | Expansion and Rebuilding** | New Projects | Expansion and Rebuilding** |
| **1976-1980** | **129.101** | **91.079** | **55.1** | **38.9** |
| 1978 | 28.662 | 18.758 | 57.2 | 37.4 |
| 1979 | 27.967 | 21.038 | 53.4 | 40.2 |
| 1980 | 29.024 | 23.749 | 51.9 | 42.5 |
| **1981-1985** | **159.187** | **162.267** | **46.7** | **47.6** |
| 1981 | 21.827 | 20.949 | 49.3 | 47.3 |
| 1982 | 26.990 | 26.495 | 48.6 | 47.7 |
| 1983 | 27.247 | 27.139 | 45.9 | 45.7 |
| 1984 | 34.981 | 35.495 | 47.1 | 47.7 |
| 1985 | 48.142 | 52.189 | 44.8 | 48.6 |
| **1986-1990** | | | | |
| 1986 | 54.989 | 56.965 | 46.8 | 48.5 |
| 1987 | 57.949 | 53.740 | 43.8 | 40.6 |

Note:
* The table refers to capital construction by state owned units only.

** Including investment in residential facilities.

## 6.9 CAPITAL CONSTRUCTION INVESTMENT IN PRODUCTIVE AND NONPRODUCTIVE PROJECTS*

| Year | Investment (1,000,000,000 yuan) | | | Percentage of Total Investment | | |
|---|---|---|---|---|---|---|
| | Productive Projects** | Non-productive Projects** | | Productive Projects** | Non-productive Projects** | |
| | | *Total* | *Residential Buildings* | | *Total* | *Residential Buildings* |
| **1976-1980** | **172.994** | **61.223** | **27.729** | **73.9** | **26.1** | **11.8** |
| 1978 | 39.624 | 10.475 | 3.921 | 79.1 | 20.9 | 7.8 |
| 1979 | 36.514 | 15.834 | 7.728 | 69.8 | 30.2 | 14.8 |
| 1980 | 35.928 | 19.961 | 11.166 | 64.3 | 35.7 | 20.0 |
| **1981-1985** | **195.651** | **145.358** | **72.699** | **57.4** | **42.6** | **21.3** |
| 1981 | 25.243 | 19.048 | 11.119 | 57.0 | 43.0 | 25.1 |
| 1982 | 30.290 | 25.263 | 14.105 | 54.5 | 45.5 | 25.4 |
| 1983 | 34.644 | 24.769 | 12.507 | 58.3 | 41.7 | 21.1 |
| 1984 | 44.340 | 29.975 | 13.450 | 59.7 | 40.3 | 18.1 |
| 1985 | 61.134 | 46.303 | 21.518 | 56.9 | 43.1 | 20.0 |
| **1986-1990** | | | | | | |
| 1986 | 71.215 | 46.395 | 18.941 | 60.6 | 39.4 | 16.1 |
| 1987 | 87.223 | 45.186 | 17.738 | 65.9 | 34.1 | 13.4 |

Note:
* The table refers to capital construction by state owned units only.

** Productive projects are projects for material production or projects directly serving material production. Non productive projects refer to projects of residential buildings, scientific research institutions, cultural, educational and healthcare facilities, public services, and other administrative agencies that serve the needs of people.

## 6.10 CAPITAL CONSTRUCTION INVESTMENT IN AGRICULTURE, AND LIGHT AND HEAVY INDUSTRY*

| Year | Investment (1,000,000,000 yuan) | | | Percentage of Total Investment (Total Investment=100) | | |
|---|---|---|---|---|---|---|
| | Agriculture | Light Industry | Heavy Industry | Agriculture | Light Industry | Heavy Industry |
| ***1976-1980*** | **24.608** | **15.625** | **107.546** | **10.5** | **6.7** | **45.9** |
| *1978* | 5.334 | 2.930 | 24.386 | 10.6 | 5.8 | 48.7 |
| *1979* | 5.792 | 3.060 | 22.625 | 11.1 | 5.8 | 43.2 |
| *1980* | 5.203 | 5.089 | 22.472 | 9.3 | 9.1 | 40.2 |
| ***1981-1985*** | **17.284** | **23.445** | **131.252** | **5.1** | **6.9** | **38.5** |
| *1981* | 2.921 | 4.338 | 17.263 | 6.6 | 9.8 | 39.0 |
| *1982* | 3.412 | 4.645 | 21.415 | 6.1 | 8.4 | 38.5 |
| *1983* | 3.545 | 3.875 | 24.353 | 6.0 | 6.5 | 41.0 |
| *1984* | 3.712 | 4.243 | 29.916 | 5.0 | 5.7 | 40.3 |
| *1985* | 3.694 | 6.344 | 38.305 | 3.4 | 5.9 | 35.7 |
| ***1986-1990*** | | | | | | |
| *1986* | 3.506 | 8.225 | 44.939 | 3.0 | 7.0 | 38.2 |
| *1987* | 6.301 | 6.747 | 61.458 | 4.8 | 5.1 | 46.4 |

**Note:**
*** The table refers to capital construction by state owned units.**

## 6.11 CAPITAL CONSTRUCTION INVESTMENT IN ENERGY INDUSTRY*, AND TRANSPORTATION, POSTAL AND TELECOMMUNICATION SERVICES**

| Year | Investment (1,000,000,000 yuan) | | Percentage of Total Investment (Total Investment =100) | |
|---|---|---|---|---|
| | Energy Industry | Transportation, Postal and Telecommunication Services | Energy Industry | Transportation, Postal and Telecommunication Services |
| ***1976-1980*** | **48.990** | **30.245** | **20.9** | **12.9** |
| *1978* | 11.470 | 6.804 | 22.9 | 13.6 |
| *1979* | 11.098 | 6.409 | 21.2 | 12.2 |
| *1980* | 11.565 | 6.234 | 20.7 | 11.2 |
| ***1981-1985*** | **69.403** | **45.513** | **20.4** | **13.3** |
| *1981* | 9.464 | 4.047 | 21.4 | 9.1 |
| *1982* | 10.223 | 5.721 | 18.4 | 10.3 |
| *1983* | 12.760 | 7.804 | 21.5 | 13.1 |
| *1984* | 16.596 | 10.846 | 22.3 | 14.6 |
| *1985* | 20.360 | 17.095 | 19.0 | 15.9 |
| ***1986-1990*** | | | | |
| *1986* | 26.510 | 18.081 | 22.5 | 15.4 |
| *1987* | 30.865 | 20.375 | 23.3 | 15.4 |

**Note:**
*** Energy industry includes coal industry, power generation industry, petroleum industry and coking chemical industry.**

**** The table refers to capital construction by state owned units.**

## 6.12 NEWLY INCREASED PRODUCTION CAPACITY THROUGH CAPITAL CONSTRUCTION*

| Year | Coal Mining (1,000,000 tn) | Capacity of Power Generating Equipment (1,000,000 Kw) | Petroleum Drilling (1,000,000 tn) | Iron Ore Mining (1,000,000 tn) | Steel Smelting (1,000,000 tn) | Iron Smelting (1,000,000 tn) | Synthetic Ammonia (1,000,000 tn) |
|---|---|---|---|---|---|---|---|
| ***1976-1980*** | **64.93** | **19.291** | **39.753** | **20.970** | **5.830** | **3.600** | **5.925** |
| *1978* | 11.51 | 5.048 | 9.996 | 1.150 | 1.125 | 1.438 | 0.958 |
| *1979* | 13.93 | 4.652 | 8.000 | 4.620 | 2.105 | 1.530 | 0.939 |
| *1980* | 8.29 | 2.871 | 5.747 | 2.740 | 0.708 | | 0.330 |
| ***1981-1985*** | **81.27** | **20.226** | **50.131** | **13.410** | **0.300** | **0.930** | **1.726** |
| *1981* | 13.73 | 2.640 | 5.189 | 4.750 | 0.000 | | 0.375 |
| *1982* | 8.20 | 2.943 | 6.365 | 3.100 | 0.180 | 0.250 | 0.725 |
| *1983* | 18.52 | 4.491 | 8.108 | 0.300 | 0.060 | 0.170 | 0.117 |
| *1984* | 24.35 | 3.779 | 13.097 | 0.460 | 0.000 | | 0.077 |
| *1985* | 16.47 | 6.376 | 17.372 | 4.800 | 0.060 | 0.510 | 0.432 |
| ***1986-1990*** | | | | | | | |
| *1986* | 21.05 | 6.638 | 15.485 | 9.925 | 3.855 | 3.084 | 0.472 |
| *1987* | 17.11 | 0.810 | 16.315 | 4.000 | 0.025 | 0.450 | 0.042 |

### NEWLY INCREASED PRODUCTION CAPACITY THROUGH CAPITAL CONSTRUCTION
(continued)

| Year | Chemical Fertilizer (1,000,000 tn) | Cement (1,000,000 tn) | Felling and Transport of Timber (1,000,000 cu. m) | Sulphuric Acid (1,000,000 tn) | Soda Ash (1,000,000 tn) | Cotton Yarn Weaving Spindles (1,000,000) | Chemical Fibers (1,000,000 tn) |
|---|---|---|---|---|---|---|---|
| ***1976-1980*** | **4.7384** | **11.196** | **3.614** | **0.881** | **0.189** | **2.012** | **0.2535** |
| *1978* | 0.8353 | 1.891 | 0.778 | 0.220 | 0.005 | 0.272 | 0.0304 |
| *1979* | 0.8221 | 2.738 | 0.892 | 0.111 | 0.115 | 0.625 | 0.0733 |
| *1980* | 0.2794 | 2.888 | 0.498 | 0.080 | 0.044 | 0.761 | 0.0605 |
| ***1981-1985*** | **1.6202** | **13.274** | **2.027** | **0.249** | **0.096** | **1.569** | **0.3204** |
| *1981* | 0.3228 | 1.544 | 0.298 | 0.035 | | 0.510 | 0.0795 |
| *1982* | 0.6532 | 2.368 | 0.333 | 0.097 | | 0.510 | 0.0291 |
| *1983* | 0.0825 | 3.457 | 0.447 | | | 0.312 | 0.0513 |
| *1984* | 0.1257 | 4.740 | 0.549 | 0.045 | 0.020 | 0.131 | 0.0260 |
| *1985* | 0.4360 | 1.165 | 0.400 | 0.072 | 0.076 | 0.106 | 0.1345 |
| ***1986-1990*** | | | | | | | |
| *1986* | 0.4427 | 5.135 | 0.438 | 0.346 | 0.255 | 0.128 | 0.1542 |
| *1987* | 0.3040 | 2.974 | 0.230 | | 0.264 | | |

**Note:**
*** The table refers to capital construction by state owned units.**

## *NEWLY INCREASED PRODUCTION CAPACITY THROUGH CAPITAL CONSTRUCTION*
(continued)

| Year | Machine Refined Sugar (1,000,000 tn) | Machine Made Paper and Paper Boards (1,000,000 tn) | Highway Construction (km) | Newly Operational Railway Lines (km) | Length of Electrified Railway (km) | New Double Line Railways (km) | Handling Capacity of Ports (1,000,000 tons) |
|---|---|---|---|---|---|---|---|
| ***1976-1980*** | **0.716** | **0.346** | **40344** | **3776** | **1000.0** | **878.0** | **69.80** |
| *1978* | 0.121 | 0.035 | 10578 | 1296 | | 95.5 | 16.57 |
| *1979* | 0.226 | 0.112 | 4956 | 289 | | 332.8 | 8.36 |
| *1980* | 0.114 | 0.108 | 3036 | 1008 | 643.0 | 158.4 | 6.83 |
| ***1981-1985*** | **1.581** | **0.329** | **8352** | **2238** | **2505.6** | **1580.6** | **99.83** |
| *1981* | 0.251 | 0.067 | 1554 | | | 95.9 | 3.36 |
| *1982* | 0.345 | 0.059 | 751 | 31 | 116.0 | 257.7 | 20.00 |
| *1983* | 0.331 | 0.092 | 1462 | 601 | 544.3 | 411.0 | 17.73 |
| *1984* | 0.366 | 0.078 | 1443 | 1247 | 694.9 | 575.3 | 9.18 |
| *1985* | 0.288 | 0.033 | 3142 | 359 | 1150.4 | 240.7 | 49.56 |
| ***1986-1990*** | | | | | | | |
| *1986* | 0.124 | 0.011 | 4073 | 910 | 285.7 | 459.9 | 46.46 |
| *1987* | 0.560 | 0.072 | 226 | 272 | 204.3 | 432.5 | 98.40 |

# 6.13 NEWLY INCREASED FIXED ASSETS AND FLOOR SPACE OF BUILDINGS COMPLETED OR UNDER CONSTRUCTION THROUGH CAPITAL CONSTRUCTION BY REGION 1987

| Region | Newly Increased Fixed Assets (1,000,000,000 yuan) | Floor Space of Buildings under Construction (1,000,000 sq. meters) | | Floor Space of Buildings Completed (1,000,000 sq. meters) | |
|---|---|---|---|---|---|
| | | Total | Residential Buildings | Total | Residential Buildings |
| ***National Total*** | **88.642** | **296.6822** | **12.0574** | **142.1046** | **64.965** |
| *Beijing* | 4.184 | 15.7600 | 69.00 | 4.9700 | 2.470 |
| *Tianjin* | 1.800 | 4.3845 | 13.31 | 1.9629 | 0.735 |
| *Hebei* | 3.069 | 12.0220 | 44.10 | 6.0260 | 2.495 |
| *Shanxi* | 4.261 | 10.6140 | 38.54 | 4.2670 | 1.822 |
| *Inner Mongolia* | 1.823 | 5.0800 | 16.60 | 2.9800 | 1.000 |
| *Liaoning* | 5.358 | 15.8320 | 72.22 | 7.1790 | 3.628 |
| *Jilin* | 2.277 | 6.1169 | 25.02 | 3.1724 | 1.499 |
| *Heilongjiang* | 4.723 | 10.2600 | 40.50 | 6.0300 | 3.030 |
| *Shanghai* | 3.446 | 16.3800 | 81.05 | 5.6380 | 2.779 |
| *Jiangsu* | 4.560 | 13.7618 | 58.45 | 7.8347 | 3.181 |
| *Zhejiang* | 2.359 | 10.6475 | 38.24 | 5.4638 | 2.326 |
| *Anhui* | 2.068 | 7.6452 | 27.07 | 3.4277 | 1.367 |
| *Fujian* | 2.191 | 8.5589 | 28.47 | 3.8029 | 1.406 |
| *Jiangxi* | 0.832 | 5.9106 | 20.85 | 2.9838 | 1.243 |
| *Shandong* | 5.249 | 11.7790 | 42.14 | 6.1490 | 2.625 |
| *Henan* | 2.477 | 11.1400 | 47.50 | 6.2100 | 2.900 |
| *Hubei* | 3.637 | 16.0600 | 71.00 | 8.4100 | 4.210 |
| *Hunan* | 2.182 | 11.0490 | 46.07 | 5.9646 | 2.993 |
| *Guangdong* | 9.463 | 22.1547 | 77.91 | 9.9054 | 3.852 |
| *Guangxi* | 1.130 | 7.9938 | 42.61 | 5.2456 | 3.177 |
| *Sichuan* | 4.625 | 23.7213 | 115.33 | 12.0104 | 6.188 |
| *Guizhou* | 0.809 | 5.4400 | 20.40 | 2.6200 | 1.130 |
| *Yunnan* | 1.330 | 6.1930 | 26.94 | 3.3950 | 1.638 |
| *Tibet* | 0.073 | 0.6526 | 3.07 | 0.5024 | 0.193 |
| *Shaanxi* | 2.114 | 9.6600 | 43.10 | 4.2000 | 2.100 |
| *Gansu* | 1.225 | 9.8640 | 23.12 | 2.5090 | 1.005 |
| *Qinghai* | 1.468 | 2.7044 | 11.10 | 1.6701 | 0.613 |
| *Ningxia* | 0.678 | 1.8300 | 5.00 | 1.0800 | 0.380 |
| *Xinjiang* | 1.538 | 6.0920 | 23.13 | 3.2760 | 1.183 |
| *Non-Region Specific* | 7.691 | 7.3750 | 33.91 | 3.2189 | 1.798 |

## 6.14 *TECHNICAL UPDATING AND TRANSFORMATION INVESTMENT BY ADMINISTATIVE LEVEL AND SECTOR*

(1,000,000,000 yuan)

| Item | 1981 | 1985 | 1986 | 1987 |
|---|---|---|---|---|
| **Total** | **19.530** | **4.4914** | **61.921** | **74.263** |
| **Administrative Level** | | | | |
| Central Government Projects | 6.088 | 10.482 | 14.867 | 17.909 |
| Local Projects | 13.442 | 34.432 | 47.054 | 56.355 |
| **Sector** | | | | |
| Industry | 14.204 | 33.764 | 45.239 | 54.233 |
| *Electronics Industry* | *0.670 | 2.001 | 2.199 | 1.762 |
| *Ship-building Industry* | | 0.178 | 0.144 | 0.225 |
| *Metallurgical Industry* | **1.654 | 4.117 | 6.014 | 7.685 |
| *Non-ferrous Metal Industry* | | 1.027 | 1.690 | 2.074 |
| *Coal Industry* | 1.301 | 2.531 | 2.969 | 3.117 |
| *Pertroleum Industry* | 2.669 | 1.184 | 1.362 | 1.542 |
| *Pertrol Chemistry Industry* | | 0.949 | 1.240 | 1.676 |
| *Chemical Industry* | 1.331 | 2.881 | 3.862 | 5.253 |
| *Power Industry* | 0.738 | 1.351 | 1.775 | 2.067 |
| *Machine-building Industry* | ***0.641 | 3.176 | 4.343 | 5.757 |
| *Automobile Industry* | | 0.556 | 0.588 | 0.850 |
| *Building Material Industy* | 0.419 | 1.756 | 2.519 | 2.690 |
| *Textile Industry* | 1.777 | 3.840 | 5.377 | 6.680 |
| *Light Industry* | 2.284 | 5.597 | 7.486 | 8.962 |
| *Other Industry* | 0.720 | 2.620 | 3.671 | 3.893 |
| Agriculture, Forestry, Water Conservancy, Meteorology | 0.573 | 1.028 | 1.396 | 1.504 |
| Transportation, Postal, and Telecommunication Services | 2.332 | 4.213 | 5.855 | 7.162 |
| Commerce and Foreign Trade | 0.927 | 2.240 | 3.223 | 4.305 |
| Urban Construction | 0.770 | 2.181 | 4.039 | 4.029 |
| Others | 0.724 | 1.488 | 2.169 | 3.030 |

**Note:**
*** This number includes two industries: the electronics industry and the ship-building industry.**

**** This number includes two industries: the metallurigical industry and the non-ferrous metal industry.**

***** This number includes two industries: the engineering industry and the automobile industry.**

## 6.15 TECHNICAL UPDATING AND TRANSFORMATION INVESTMENT BY PURPOSE, USE OF FUNDS, AND TYPE OF CONSTRUCTION

(1,000,000,000 yuan)

| Item | 1981 | 1985 | 1986 | 1987 |
|---|---|---|---|---|
| **Total** | **19.530** | **44.914** | **61.921** | **74.263** |
| **Purpose** | | | | |
| Productive Construction | 16.072 | 38.070 | 51.450 | 61.356 |
| *Energy Saving* | 0.779 | 1.737 | 2.275 | 282.9 |
| *Expanding Production* | | 16.035 | 21.120 | 25.623 |
| *Increasing Variety* | *9.320 | 6.130 | 9.788 | 10.598 |
| *Improving Quality* | 0.708 | 2.658 | 3.916 | 4.200 |
| *Treatment of Waste Water, Gas and Industrial Waste* | 0.298 | 0.908 | 1.178 | 1.387 |
| *Others* | 4.967 | 10.602 | 13.173 | 16.719 |
| Non-Productive Construction | 3.458 | 6.844 | 10.471 | 12.908 |
| *Residential Buildings* | 1.781 | 2.525 | 4.304 | 6.189 |
| **Use of Funds** | | | | |
| Construction and Installation of Projects | 11.377 | 19.623 | 26.779 | 33.798 |
| Purchase of Equipment, Tools and Instruments | 7.615 | 22.494 | 30.858 | 34.039 |
| *Purchase of Updated Equipment* | | 5.836 | 8.885 | 9.806 |
| **Construction** | | | | |
| New Construction | 1.975 | 2.306 | 3.626 | 4.548 |
| Expansion | 7.501 | 19.445 | 27.459 | 32.400 |
| Rebuilding | 8.902 | 19.116 | 26.184 | 30.658 |

**Note:**
*** This number includes two items: increasing variety and improving quality.**

## 6.16 *TECHNICAL UPDATING AND TRANSFORMATION INVESTMENT BY REGION*

(1,000,000,000 yuan)

| Region | 1981 | 1985 | 1986 | 1987 |
|---|---|---|---|---|
| ***National Total*** | **19.530** | **44.914** | **61.921** | **74.263** |
| *Beijing* | 0.680 | 2.118 | 3.288 | 4.299 |
| *Tianjin* | 0.745 | 2.062 | 3.151 | 2.776 |
| *Hebei* | 1.010 | 1.722 | 2.487 | 3.108 |
| *Shanxi* | 0.539 | 1.431 | 1.754 | 2.081 |
| *Inner Mongolia* | 0.343 | 0.741 | 1.129 | 1.219 |
| *Liaoning* | 1.932 | 4.252 | 6.280 | 7.558 |
| *Jilin* | 0.789 | 1.215 | 1.592 | 2.061 |
| *Heilongjiang* | 1.457 | 1.495 | 2.302 | 2.751 |
| *Shanghai* | 1.114 | 3.561 | 4.683 | 5.737 |
| *Jiangsu* | 0.825 | 2.473 | 3.404 | 4.211 |
| *Zhejiang* | 0.603 | 1.117 | 1.676 | 2.153 |
| *Anhui* | 0.402 | 1.260 | 1.864 | 2.063 |
| *Fujian* | 0.358 | 0.849 | 1.046 | 1.255 |
| *Jiangxi* | 0.433 | 0.772 | 1.169 | 1.275 |
| *Shandong* | 1.097 | 2.120 | 2.643 | 3.733 |
| *Henan* | 0.764 | 1.630 | 2.242 | 2.851 |
| *Hubei* | 0.619 | 1.822 | 2.223 | 2.675 |
| *Hunan* | 0.562 | 1.603 | 2.003 | 2.456 |
| *Guangdong* | 0.635 | 3.435 | 4.525 | 4.821 |
| *Guangxi* | 0.336 | 0.712 | 1.084 | 1.648 |
| *Sichuan* | 1.380 | 3.109 | 3.924 | 4.870 |
| *Guizhou* | 0.276 | 0.763 | 0.907 | 0.950 |
| *Yunnan* | 0.469 | 0.977 | 1.290 | 1.425 |
| *Tibet* | 0.047 | 0.026 | 0.016 | 0.016 |
| *Shaanxi* | 0.409 | 1.173 | 1.593 | 1.933 |
| *Gansu* | 0.367 | 0.778 | 1.076 | 1.305 |
| *Qinghai* | 0.055 | 0.142 | 0.170 | 0.191 |
| *Ningxia* | 0.131 | 0.259 | 0.380 | 0.388 |
| *Xinjiang* | 0.551 | 0.611 | 0.849 | 0.904 |
| *Non-Region Specific* | 0.649 | 0.665 | 1.162 | 1.553 |

## 6.17 *NEWLY INCREASED FIXED ASSETS AND FLOOR SPACE OF BUILDINGS UNDER CONSTRUCTION THROUGH TECHNICAL UPDATING AND TRANSFORMATION INVESTMENT BY REGION 1987*

| Region | Newly Increased Fixed Assets (1,000,000,000 yuan) | Floor Space of Buildings under Construction (1,000,000 sq. meters) | | Floor Space of Buildings Completed (1,000,000 sq. meters) | |
|---|---|---|---|---|---|
| | | Total | Residential Buildings | Total | Residential Buildings |
| ***National Total*** | **54.856** | **120.7352** | **41.1722** | **61.0912** | **21.3376** |
| *Beijing* | 3.215 | 4.6690 | 2.1946 | 2.2161 | 0.9916 |
| *Tianjin* | 2.495 | 3.6666 | 1.8249 | 1.8981 | 0.9819 |
| *Hebei* | 2.286 | 4.3501 | 1.0330 | 1.9409 | 0.3496 |
| *Shanxi* | 1.473 | 3.2070 | 1.3280 | 1.6300 | 0.6820 |
| *Inner Mongolia* | 1.070 | 2.2600 | 0.8200 | 1.3600 | 0.5500 |
| *Liaoning* | 5.937 | 10.8090 | 4.5430 | 5.9050 | 2.6520 |
| *Jilin* | 1.616 | 1.9290 | 0.3567 | 1.0719 | 0.2755 |
| *Heilongjiang* | 1.928 | 3.0689 | 0.8883 | 1.6105 | 0.5087 |
| *Shanghai* | 3.909 | 9.3933 | 4.1320 | 4.0946 | 1.7508 |
| *Jiangsu* | 3.373 | 6.0133 | 1.2851 | 3.0942 | 0.6988 |
| *Zhejiang* | 1.636 | 4.0738 | 0.9250 | 1.9156 | 0.3045 |
| *Anhui* | 1.759 | 3.3983 | 0.7808 | 1.7103 | 0.3919 |
| *Fujian* | 0.934 | 2.1387 | 1.0097 | 1.8894 | 0.6130 |
| *Jiangxi* | 0.750 | 3.3071 | 1.0097 | 1.8894 | 0.6130 |
| *Shandong* | 2.860 | 5.8000 | 1.8168 | 3.0835 | 0.9629 |
| *Henan* | 2.001 | 5.5560 | 1.8554 | 2.8123 | 0.9435 |
| *Hubei* | 1.676 | 3.7744 | 0.7844 | 1.7884 | 0.4182 |
| *Hunan* | 1.819 | 4.5528 | 1.4224 | 2.4082 | 0.8316 |
| *Guangdong* | 3.786 | 9.0656 | 3.2643 | 4.6035 | 1.5790 |
| *Guangxi* | 1.056 | 3.5498 | 1.0844 | 1.6146 | 0.5106 |
| *Sichuan* | 3.449 | 12.0467 | 4.1222 | 6.3231 | 2.2737 |
| *Guizhou* | 0.605 | 2.1000 | 0.5800 | 1.0700 | 0.3200 |
| *Yunnan* | 0.984 | 4.0242 | 1.7483 | 2.1020 | 1.0861 |
| *Tibet* | 0.011 | 0.0099 | 0.0024 | 0.0060 | 0.0012 |
| *Shaanxi* | 1.149 | 3.0586 | 1.2469 | 1.3462 | 0.5968 |
| *Gansu* | 0.586 | 1.9614 | 0.6488 | 0.9483 | 0.3369 |
| *Qinghai* | 0.092 | 0.3296 | 0.1580 | 0.1760 | 0.0905 |
| *Ningxia* | 0.536 | 0.8800 | 0.2800 | 0.3900 | 0.1500 |
| *Xinjiang* | 0.757 | 1.6020 | 0.5470 | 0.9050 | 0.2880 |
| *Non-Region Specific* | 1.136 | 0.1401 | 0.0391 | 0.0864 | 0.0336 |

## 6.18 TECHNICAL UPDATING AND TRANSFORMATION INVESTMENT IN PROJECTS OF 30,000,000 YUAN AND OVER BY SECTOR 1987

| Sector | Number of Projects | Projects Completed and put into Use | Investment (1,000,000 yuan) | Percentage |
|---|---|---|---|---|
| **National Total** | **386** | **49** | **7473.21** | **100.0** |
| **Agriculture, Forestry, Animal Husbandry and Fishery** | 1 | | 4.08 | 0.1 |
| **Industry** | 335 | 37 | 6158.68 | 82.4 |
| Ferrous Metal Mining and Preparation | 4 | | 28.78 | 0.4 |
| Non-Ferrous Metal Mining and Preparation | 3 | | 44.63 | 0.6 |
| Salt Production | 2 | | 15.55 | 0.2 |
| Tap Water Purification and Supply | 4 | | 30.47 | 0.4 |
| Food Manufacturing | 3 | 1 | 39.44 | 0.5 |
| Beverage Manufacturing | 18 | 1 | 279.40 | 3.7 |
| Tobacco Manufacturing | 4 | 1 | 52.89 | 0.7 |
| Textile Industry | 32 | 9 | 333.36 | 4.5 |
| Paper Making and Manufactured Goods | 10 | 2 | 185.53 | 2.5 |
| Printing | 1 | | 7.26 | 0.1 |
| Manufacturing of Recreational, Educational and Sporting Goods | 2 | | 21.07 | 0.3 |
| Power Generation and Supply | 7 | 2 | 68.53 | 0.9 |
| *Electricity Supply* | 1 | | 13.95 | 0.2 |
| Petroleum Processing | 4 | | 106.43 | 1.4 |
| Coking, Gas and Coal Related Products | 4 | | 125.66 | 1.7 |
| *Coking* | 1 | | 12.29 | 0.2 |
| Chemical Industry | 39 | 4 | 762.36 | 10.2 |
| Medical and Pharmaceutical Products | 5 | | 50.76 | 0.7 |
| Chemical Fibers | 5 | 2 | 67.86 | 0.9 |
| Rubber Manufactured Goods | 1 | 1 | 3.44 | |
| Building Materials and Other Non-Metal Mineral Products | 13 | 2 | 225.75 | 3.0 |
| Smelting and Pressing of Ferrous Metals | 74 | 9 | 1988.43 | 26.6 |
| Smelting and Pressing of Non-Ferrous Metals | 18 | 1 | 436.60 | 5.8 |
| Metal Products | 7 | | 76.26 | 1.0 |
| Machine Building | 48 | 1 | 703.85 | 9.4 |
| Transportation Equipment Manufacturing | 13 | | 320.07 | 4.3 |
| Electronic Equipment Manufacturing | 2 | | 19.20 | 0.3 |
| Electronic and Telecommunication Equipment | 11 | 1 | 157.23 | 2.1 |
| Precision and Metrological Instrument | 1 | | 7.87 | 0.1 |

## *TECHNICAL UPDATING AND TRANSFORMATION INVESTMENT IN PROJECTS OF 30,000,000 YUAN AND OVER BY SECTOR 1987*
(continued)

| Sector | Number of Projects | Projects Completed and Put into Use | Investment (1,000,000 yuan) | Percentage |
|---|---|---|---|---|
| ***Transportation, Postal and Telecommunication Services*** | 34 | 7 | 741.00 | 9.9 |
| *Railway* | 22 | 1 | 551.60 | 7.4 |
| *Highway* | 1 | | 44.43 | 0.6 |
| *Pipeline* | 3 | 1 | 20.32 | 0.3 |
| *Waterway* | 3 | 2 | 15.79 | 0.2 |
| *Postal and Telecommunication Services* | 5 | 3 | 108.86 | 1.5 |
| ***Commerce, Food Service, Supply and Marketing and Storage*** | 7 | 3 | 117.76 | 1.6 |
| ***Real Estate Management, Public Utilities, Resident Services and Consultation*** | 7 | 2 | 430.23 | 5.8 |
| ***Education, Culture, Radio and Television Broadcasting and the Arts*** | 1 | | 20.03 | 0.3 |
| ***Scientific Research and Polytechnical Services*** | 1 | | 1.43 | |

## 6.19 *GROSS OUTPUT VALUE AND NET OUTPUT VALUE OF CONSTRUCTION INDUSTRY**

| Year | Gross Output Value | | Net Output Value | |
|---|---|---|---|---|
| | Total (1,000,000,000 yuan) | Index (1978 = 100) | Total (1,000,000,000 yuan) | Index (1978 = 100) |
| *1978* | 56.9 | 100.0 | 12.5 | 100.0 |
| *1979* | 64.5 | 110.7 | 13.0 | 101.8 |
| *1980* | 76.7 | 128.9 | 18.5 | 132.0 |
| *1981* | 74.7 | 121.8 | 19.3 | 134.1 |
| *1982* | 91.2 | 144.5 | 20.9 | 140.5 |
| *1983* | 105.3 | 159.6 | 25.9 | 166.2 |
| *1984* | 126.3 | 181.0 | 30.3 | 184.0 |
| *1985* | 165.6 | 217.8 | 40.9 | 221.2 |
| *1986* | 203.8 | 248.2 | 51.4 | 255.7 |
| *1987* | 244.6 | 278.5 | 61.9 | 289.5 |

**Note:**
*** The total value is calculated at the current price. The index is used on comparable prices.**

## 6.20 *PRICE INDEX OF OUTPUT VALUE AND EXPENSES OF STATE OWNED CONSTRUCTION INDUSTRY*
(previous year = 100)

| Item | 1984 | 1985 | 1986 | 1987 |
|---|---|---|---|---|
| **Output Value** | 105.8 | 109.4 | 110.0 | 108.4 |
| **Direct Expenses** | 106.7 | 110.6 | 111.3 | 108.8 |
| Labor | 105.0 | 105.5 | 108.0 | 108.1 |
| Material | 107.5 | 112.5 | 112.3 | 109.4 |
| *Steel* | 109.2 | 116.8 | 111.4 | 108.8 |
| *Timber* | 112.0 | 118.1 | 114.3 | 122.8 |
| *Cement* | 109.4 | 114.7 | 124.7 | 109.8 |
| *Local Materials** | 105.9 | 109.5 | 109.3 | 108.1 |
| *Other Materials* | 105.6 | 109.4 | 110.3 | 108.2 |
| **Other Expenses** | 101.9 | 105.2 | 105.9 | 107.3 |

**Note:**
*** Local materials refers to materials produced by local individuals and not directly distributed by central planning system.**

## 6.21 MAIN INDICATORS OF ECONOMIC EFFICIENCY OF STATE OWNED CONSTRUCTION ENTERPRISES

| Year | Overall Labor Productivity (yuan/person- year) | Floor Space of Buildings Completed per Capita (sq. m/person) | Percentage of Buildings Completed |
|---|---|---|---|
| 1980 | 4257 | 13.1 | 46.7 |
| 1981 | 4051 | 12.9 | 47.5 |
| 1982 | 4574 | 12.8 | 45.7 |
| 1983 | 5148 | 13.2 | 45.7 |
| 1984 | 6234 | 14.0 | 47.2 |
| 1985 | 7630 | 14.6 | 44.4 |
| 1986 | 8608 | 14.2 | 43.8 |
| 1987 | 9937 | 14.5 | 43.2 |

## 6.22 NUMBER OF PROJECTS BY STATE OWNED CONSTRUCTION ENTERPRISES UNDER CONTRACTUAL RESPONSBILITY SYSTEM

| Item | 1985 | 1986 | 1987 |
|---|---|---|---|
| ***Total Number of Projects*** | 153762 | 157019 | 147867 |
| ***Total Number of Projects under Contractual Responsibility System*** | 117353 | 121053 | 116617 |
| ***Type of Contractual Responsibility System*** | | | |
| *Open Bidding* | 11192 | 15634 | 17798 |
| *Cost Plus Based on Estimate of Budget* | 69625 | 71831 | 69297 |
| *Cost per Square Meter* | 13395 | 13242 | 10771 |
| *Unit Cost of Designed Production Capacity* | 1030 | 910 | 885 |
| *Package Building of Living Quarters* | 2079 | 1755 | 1464 |
| *Other* | 20032 | 17681 | 16402 |

*Section VII*

# FINANCE, TRADE AND PRICE

## 7.1 *TOTAL GOVERNMENT REVENUE AND EXPENDITURE**
(1,000,000,000 yuan)

| Year | Total Revenue | Total Expenditure | Balance |
|---|---|---|---|
| *1978* | 112.11 | 111.10 | 1.01 |
| *1979* | 110.33 | 127.39 | -17.06 |
| *1980* | 108.52 | 121.27 | -12.75 |
| *1981* | 108.95 | 111.50 | -2.55 |
| *1982* | 112.40 | 115.33 | -2.93 |
| *1983* | 124.90 | 129.25 | -4.35 |
| *1984* | 150.19 | 154.64 | -4.45 |
| *1985* | 186.64 | 184.48 | 2.16 |
| *1986* | 226.03 | 233.08 | -7.06 |
| *1987* | 234.66 | 242.69 | -8.03 |

**Note:**
*** Government revenue includes revenue from domestic and foreign loans; government expenditure includes expenditure using those loans.**

## 7.2 *TOTAL DEPOSITS AND LOANS OF RURAL CREDIT CO-OPERATIVES*
(1,000,000,000 yuan) (year-end balance)

| Item | 1980 | 1985 | 1986 | 1987 |
|---|---|---|---|---|
| ***Total Deposits*** | **27.234** | **72.490** | **96.234** | **122.521** |
| *Deposits by Collective Farming Units* | 10.548 | 7.192 | 8.387 | 8.987 |
| *Deposits by Township and Village Run Enterprises* | 2.947 | 7.213 | 9.168 | 10.470 |
| *Personal Savings Accounts of Rural Households* | 11.703 | 56.481 | 76.612 | 100.572 |
| *Other Deposits* | 2.036 | 1.604 | 2.067 | 2.492 |
| ***Total Loans*** | **8.164** | **39.996** | **56.851** | **77.135** |
| *Agricultural Loans to Collective Farming Units* | 3.454 | 4.135 | 4.464 | 6.451 |
| *Loans to Township and Village Run Enterprises* | 3.111 | 16.443 | 26.585 | 35.931 |
| *Loans to Rural Laborers* | 1.599 | 19.418 | 25.802 | 34.753 |

## 7.3 *ITEMS OF GOVERNMENT REVENUE AND EXPENDITURE*
(1,000,000,000 yuan)

| Items | 1978 | 1980 | 1984 | 1985 | 1986 | 1987 |
|---|---|---|---|---|---|---|
| **Total Revenue** | **112.11** | **108.52** | **150.19** | **186.64** | **226.03** | **234.66** |
| Revenue from State-Owned Enterprises | 57.20 | 43.52 | 27.68 | 4.38 | 4.20 | 4.16 |
| *Industrial Enterprises* | 44.04 | 44.82 | 38.53 | | | |
| Total Revenue from Taxation | 51.93 | 57.17 | 94.74 | 204.08 | 209.07 | 213.48 |
| *Industrial and Commercial Tax* | 45.13 | 50.14 | 75.75 | 110.10 | 129.92 | |
| *Income Tax from State Owned Units* | | | | 51.38 | 52.37 | |
| *Customs Tax* | | | | 20.52 | 15.16 | |
| *Agricultural Tax* | 2.84 | 2.77 | 3.46 | 4.21 | 4.45 | |
| **Total Expenditure** | **111.10** | **121.27** | **154.64** | **184.48** | **233.08** | **242.69** |
| Funds Allocated for Capital Construction | 45.19 | 41.94 | 48.89 | 58.38 | 67.18 | 61.20 |
| Funds for Tapping Potential, Reforming Enterprises and Trial Manufacturing | | | 11.18 | 10.34 | 12.99 | 12.14 |
| Increasing Circulating Funds for Enterprises | 6.66 | 3.67 | 1.00 | 1.43 | 0.99 | |
| Culture, Education, Science, Public Health | 11.27 | 15.63 | 26.32 | 31.67 | 37.99 | 40.56 |
| *Education* | 6.56 | 9.42 | 14.82 | 18.42 | 21.43 | |
| National Defence and Combat Readiness | 16.78 | 19.38 | 18.08 | 19.15 | 20.08 | 20.98 |
| Administration | 4.90 | 6.68 | 13.73 | 14.36 | 16.80 | 17.97 |

## 7.4 *GOLD AND FOREIGN EXCHANGE RESERVES*
(year-end balance)

| Item | 1980 | 1985 | 1986 | 1987 |
|---|---|---|---|---|
| *Gold (1,000,000 oz.)* | 12.80 | 12.67 | 12.67 | 12.67 |
| *Foreign Exchange (1,000,000,000 U.S. dollars)* | 2.262 | 11.913 | 10.514 | 15.236 |

# 7.5 SOURCES AND USES OF FUNDS OF THE NATIONAL BANKING SYSTEM

(1,000,000,000 yuan) (total year-end balance)

| Item | 1980 | 1985 | 1986 | 1987 |
|---|---|---|---|---|
| **Total Source of Funds** | **262.426** | **643.087** | **820.640** | **997.617** |
| All Deposits | 165.864 | 427.303 | 538.152 | 657.205 |
| *Enterprise Deposits* | 57.309 | 207.153 | 264.301 | 312.555 |
| *Treasury Deposits* | 16.202 | 36.837 | 31.145 | 30.698 |
| *Capital Construction Deposits* | 17.175 | | | |
| *Deposits of Government Agencies and Organization* | 22.945 | 32.576 | 39.597 | 44.920 |
| *Urban Savings Deposits* | 28.249 | 105.781 | 147.145 | 206.402 |
| *Rural Savings Deposits* | 23.984 | 44.956 | 55.964 | 62.630 |
| Loan from International Monetary Institutions | 3.427 | 7.818 | 12.423 | 18.563 |
| Currency in Circulation | 34.620 | 98.783 | 121.836 | 145.448 |
| Funds of Banks | 47.733 | 77.782 | 86.165 | 94.002 |
| Current Balance of Profit and Loss Accounts | 1.972 | 7.030 | 7.937 | 12.156 |
| Others | 8.810 | 24.371 | 54.127 | 70.243 |
| **Total Use of Funds** | **262.426** | **643.087** | **820.640** | **997.617** |
| All Loans | 241.430 | 590.551 | 759.083 | 903.235 |
| *Loans to Industrial Enterprises* | 43.158 | 116.508 | 165.311 | 204.361 |
| *Loans to Industrial Supply and Marketing Enterprises and Storage Department* | 23.603 | 38.083 | 47.705 | 49.347 |
| *Commercial Loans* | 143.702 | 264.930 | 308.941 | 350.614 |
| *Medium and Short-Term Equipment Loans* | 5.550 | | | |
| *Loans to Building Enterprises* | | 26.707 | 36.941 | 46.652 |
| *Loans tor Investment in Fixed Assets* | | 70.532 | 100.594 | 55.003 |
| *Loans to Urban Collectives and Individual Industrial and Commercial Units* | 7.829 | 32.128 | 42.554 | 128.675 |
| *Agricultural Loans** | 17.588 | 41.663 | 57.037 | 68.583 |
| Gold | 1.216 | 1.204 | 1.204 | 1.204 |
| Foreign Exchange | -0.847 | 9.310 | 7.712 | 18.208 |
| Assets in the International Monetary Institutions | 3.604 | 8.694 | 10.043 | 17.881 |
| Government Debt | 17.023 | 27.505 | 37.005 | 51.496 |
| Other Uses of Funds | | 5.648 | 5.593 | 5.593 |

**Note:**
*** The agricultural loans include loans as advance payments for purchase of agricultural products, loans to state farms and loans to townships and villages.**

## 7.6 *TOTAL VALUE OF RETAIL SALES BY TYPE OF COMMODITY AND LOCATION*

(1,000,000,000 yuan)

| Year | Total Value of Retail Sales | Type of Commodity | | Location | |
|---|---|---|---|---|---|
| | | Consumer Goods | Means of Agricultural Production | | |
| *1978* | 155.86 | 126.49 | 29.37 | 74.82 | 81.04 |
| *1979* | 180.00 | 147.60 | 32.40 | 81.52 | 98.48 |
| *1980* | 214.00 | 179.40 | 34.60 | 95.03 | 118.97 |
| *1981* | 235.00 | 200.25 | 34.75 | 102.60 | 132.40 |
| *1982* | 257.00 | 218.15 | 38.85 | 109.00 | 148.00 |
| *1983* | 284.94 | 242.61 | 42.33 | 117.94 | 167.00 |
| *1984* | 337.64 | 289.92 | 47.72 | 137.71 | 199.93 |
| *1985* | 430.50 | 380.14 | 50.36 | 178.80 | 251.70 |
| *1986* | 495.00 | 437.40 | 57.60 | 209.40 | 285.60 |
| *1987* | 582.00 | 511.50 | 70.50 | 247.00 | 335.00 |

**Note:**
*** This table includes the value of retail sales by rural laborers to non-agricultural residents.**

## 7.7 *TOTAL VALUE OF RETAIL SALES BY TYPES OF SELLERS**

(1,000,000,000 yuan)

| Year | Total Value of Retail Sales | Ownership | | | | Retail Sales by Rural to Urban Residents |
|---|---|---|---|---|---|---|
| | | *State Owned Units* | *Collective Owned Units*** | *Joint Owned Units**** | *Individual Owned Units* | |
| *1978* | 155.86 | 141.01 | 11.53 | | 0.21 | 3.11 |
| *1979* | 180.00 | 158.89 | 15.93 | | 0.43 | 4.75 |
| *1980* | 214.00 | 179.78 | 25.78 | 0.04 | 1.50 | 6.90 |
| *1981* | 235.00 | 188.06 | 34.15 | 0.11 | 3.74 | 8.94 |
| *1982* | 257.00 | 196.86 | 41.44 | 0.16 | 7.46 | 11.08 |
| *1983* | 284.94 | 205.44 | 47.39 | 0.36 | 18.45 | 13.30 |
| *1984* | 337.64 | 153.79 | 133.72 | 0.76 | 32.37 | 17.00 |
| *1985* | 430.50 | 174.00 | 160.03 | 1.27 | 66.10 | 29.10 |
| *1986* | 495.00 | 195.10 | 180.40 | 1.52 | 80.48 | 37.50 |
| *1987* | 582.00 | 224.90 | 207.96 | 1.88 | 101.16 | 46.10 |

**Note:**
*** This table includes the value of retail sales by rural laborers to non-agricultural residents.**

**** The 1984 and 1985 retail sales of collective owned units include those of supply and marketing co-operatives, which were included under state owned units previously.**

***** Before 1958, joint owned enterprises were those under joint state-private ownerships. Since 1981 they include various forms of jointly owned enterprises and joint Chinese-foreign enterprises.**

## 7.8 *TOTAL VALUE OF RETAIL SALES BY REGION 1987*

(1,000,000,000 yuan)

| | Total Value of Retail Sales | Consumer Goods | Means of Agricultural Production |
|---|---|---|---|
| ***National Total*** | **582.00** | **511.50** | **70.50** |
| *Beijing* | 17.66 | 16.90 | 0.76 |
| *Tianjin* | 10.25 | 9.55 | 0.70 |
| *Hebei* | 27.13 | 22.34 | 4.79 |
| *Shanxi* | 12.84 | 11.36 | 1.48 |
| *Inner Mongolia* | 10.54 | 9.63 | 0.91 |
| *Liaoning* | 31.30 | 28.59 | 2.71 |
| *Jilin* | 16.28 | 14.55 | 1.73 |
| *Heilongjiang* | 23.64 | 21.42 | 2.22 |
| *Shanghai* | 23.87 | 22.52 | 1.35 |
| *Jiangsu* | 42.20 | 36.07 | 6.13 |
| *Zhejiang* | 27.94 | 24.26 | 3.68 |
| *Anhui* | 20.10 | 16.56 | 3.54 |
| *Fujian* | 14.21 | 12.60 | 1.61 |
| *Jiangxi* | 12.61 | 10.88 | 1.73 |
| *Shandong* | 37.47 | 30.07 | 7.40 |
| *Henan* | 27.85 | 22.53 | 5.32 |
| *Hubei* | 26.87 | 23.41 | 3.46 |
| *Hunan* | 24.29 | 20.95 | 3.34 |
| *Guangdong* | 47.95 | 43.20 | 4.75 |
| *Guangxi* | 13.66 | 11.79 | 1.87 |
| *Sichuan* | 37.20 | 32.50 | 4.70 |
| *Guizhou* | 7.32 | 6.64 | 0.68 |
| *Yunnan* | 11.33 | 10.25 | 1.08 |
| *Tibet* | | | |
| *Shaanxi* | 12.10 | 10.62 | 1.48 |
| *Gansu* | 8.00 | 7.21 | 0.79 |
| *Qinghai* | 2.28 | 2.14 | 0.14 |
| *Ningxia* | 2.00 | 1.72 | 0.28 |
| *Xinjiang* | 7.84 | 7.24 | 0.60 |

**Note:**
*** This table includes the value of retail sales from agricultural to non-agricultural residents.**

## 7.9 *TOTAL VALUE OF RETAIL SALES OF CONSUMER GOODS**

| | 1978 | 1980 | 1985 | 1986 | 1987 |
|---|---|---|---|---|---|
| ***Value of Retail Sales* (1,000,000,000 yuan)** | **126.49** | **179.40** | **380.14** | **437.40** | **511.50** |
| *Food* | 65.58 | 91.81 | 200.35 | 234.01 | 276.08 |
| *Clothes* | 27.85 | 41.37 | 71.74 | 77.26 | 86.93 |
| *Daily Articles* | 27.53 | 39.42 | 97.22 | 113.56 | 134.24 |
| *Fuel* | 5.53 | 6.80 | 10.83 | 12.57 | 14.25 |
| ***Percentage*** | **100.0** | **100.0** | **100.0** | **100.0** | **100.0** |
| *Food* | 51.8 | 51.2 | 52.7 | 53.5 | 54.0 |
| *Clothes* | 22.0 | 23.0 | 18.9 | 17.7 | 17.0 |
| *Daily Articles* | 21.8 | 22.0 | 25.6 | 25.9 | 26.2 |
| *Fuel* | 4.4 | 3.8 | 2.8 | 2.9 | 2.8 |

**Note:**
*** This table includes the value of retail sales by farmers to non-agricultural residents.**

## 7.10 *URBAN AND RURAL FREE MARKET TRADE*

| Item | 1978 | 1980 | 1985 | 1986 | 1987 |
|---|---|---|---|---|---|
| **Number of Free Markets** | **33302** | **40809** | **61337** | **67610** | **69683** |
| Urban | | 2919 | 8013 | 9701 | 10908 |
| Rural | 33302 | 37890 | 53324 | 57909 | 58775 |
| **Value of Free Market Trade (1,000,000,000 yuan)*** | **12.50** | **23.50** | **70.50** | **90.65** | **115.78** |
| Urban | | 2.40 | 18.10 | 24.44 | 34.70 |
| Rural | 12.50 | 21.10 | 52.40 | 66.21 | 81.00 |
| *Grain and Vegatable Oil* | 2.01 | 3.44 | 4.96 | 7.12 | 8.47 |
| *Meat, Poultry and Eggs* | 2.12 | 4.21 | 14.01 | 24.68 | 32.03 |
| *Aquatic Products* | 0.52 | 0.93 | 3.32 | 6.44 | 8.54 |
| *Vegetables* | 1.43 | 2.15 | 4.88 | 9.69 | 13.11 |
| *Dried and Fresh Fruits* | 0.40 | 0.75 | 2.55 | 5.93 | 8.31 |
| *Forage and Farm Tools* | 1.04 | 0.71 | 1.39 | 1.52 | 1.58 |
| *Large Domestic Animals* | 2.09 | 2.65 | 3.26 | 3.11 | 3.26 |

**Note:**
*** Value of free market trade includes value of transaction within free markets and sales by peasants to non-agricultural residents.**

# 7.11 VOLUME OF RETAIL SALES OF CONSUMER GOODS

| Product | | 1978 | 1980 | 1985 | 1986 | 1987 |
|---|---|---|---|---|---|---|
| **Grain** | 1,000,000 tons | 47.500 | 54.970 | 90.116 | 94.409 | 94.598 |
| **Vegetable Oil** | 1,000,000 tons | 0.875 | 1.260 | 3.491 | 3.645 | 3.780 |
| **Pork** | 1,000,000 tons | 4.675 | 7.045 | 9.164 | 9.591 | 9.800 |
| **Fresh Eggs** | 1,000,000 tons | 0.459 | 0.837 | 1.927 | 2.243 | 2.345 |
| **Aquatic Products** | 1,000,000 tons | 2.190 | 2.029 | 2.684 | 3.147 | 3.391 |
| **Sugar** | 1,000,000 tons | 3.156 | 3.635 | 5.725 | 6.303 | 6.996 |
| **Cigarettes** | 1,000,000 tons | 11.798 | 15.932 | 25.483 | 27.267 | 29.312 |
| **Liquor, Wine and Beer** | 1,000,000 tons | 2.460 | 3.342 | 7.996 | 9.567 | 10.619 |
| **Tea** | 1,000,000 tons | 0.110 | 0.166 | 0.253 | 0.275 | 0.289 |
| **Various Types of Cloth** | 1,000,000,000 meters | 7.69 | 9.84 | 12.12 | 11.78 | 12.13 |
| *Cotton Cloth* | 1,000,000,000 meters | 6.37 | 7.19 | 7.13 | 6.86 | 7.40 |
| *Synthetic Blend Fabrics* | 1,000,000,000 meters | * 1.32 | * 2.65 | 3.60 | 3.50 | 3.40 |
| *Chemical Fibers* | 1,000,000,000 meters | | | 1.39 | 1.42 | 1.33 |
| **Woolen Fabrics** | 1,000,000 meters | 80.979 | 142.210 | 306.400 | 295.560 | 297.037 |
| **Silk and Satin** | 1,000,000, meters | 268.026 | 441.590 | 897.588 | 902.780 | 932.572 |
| **Knitted Underwear** | 1,000,000,000 articles | 0.70 | 0.97 | 1.43 | 1.44 | 1.48 |
| **Leather Shoes** | 1,000,000,000 pairs | 0.10 | 0.17 | 0.26 | 0.28 | 0.30 |
| **Rubber Shoes** | 1,000,000,000 pairs | 0.33 | 0.42 | 0.61 | 0.63 | 0.67 |
| **Matches** | 1,000,000 boxes | 17.844 | 21.340 | 27.661 | 28.514 | 29.141 |
| **Soap** | 1,000,000 cases | 32.042 | 40.430 | 58.771 | 60.688 | 58.517 |
| **Vacuum Flasks** | 1,000,000 bottles | 81.692 | 100.510 | 152.839 | 156.749 | 153.457 |
| **Sewing Machines** | 1,000,000 units | 4.398 | 6.650 | 11.030 | 11.237 | 10.027 |
| **Aluminium Pots** | 1,000,000 | 24.798 | 29.545 | 44.582 | 43.836 | 51.288 |
| **Enamel Washbasins** | 1,000,000 | 75.270 | 87.610 | 96.367 | 98.474 | 85.672 |
| **Wrist Watches** | 1,000,000 | 13.881 | 25.340 | 61.662 | 56.834 | 54.525 |
| **Bicycles** | 1,000,000 | 8.096 | 11.860 | 31.112 | 35.093 | 35.920 |
| **Electric Fans** | 1,000,000 | 0.700 | 2.940 | 20.572 | 22.407 | 25.834 |
| **Washing Machines** | 1,000,000 | 0.002 | 0.235 | 10.981 | 12.888 | 14.151 |
| **Refrigerators** | 1,000,000 | 0.020 | 0.055 | 2.200 | 2.837 | 4.103 |
| **Television Sets** | 1,000,000 | 0.551 | 3.640 | 21.569 | 21.982 | 23.257 |
| **Recorders** | 1,000,000 | 0.080 | 1.090 | 13.243 | 15.111 | 17.804 |
| **Radio Sets** | 1,000,000 | 13.889 | 27.220 | 25.168 | 20.500 | 16.712 |
| **Machine-made Paper** | 1,000,000 tons | 0.417 | 0.418 | 0.408 | 0.410 | 0.440 |
| **Kerosene** | 1,000,000 tons | 0.974 | 0.988 | 1.556 | 1.645 | 1.614 |
| **Coal** | 1,000,000 tons | 100.630 | 125.740 | 189.780 | 223.420 | 209.134 |

**Note:**
*** These numbers are the sum of two products: synthetic blend fabrics and chemical fibers.**

## 7.12 *TOTAL PURCHASE VALUE OF FARM AND SIDELINE PRODUCTS*

(1,000,000,000 yuan)

| Item | 1978 | 1980 | 1985 | 1986 | 1987 |
|---|---|---|---|---|---|
| **Total** | **55.79** | **84.22** | **168.00** | **199.00** | **236.92** |
| **Product Source** | | | | | |
| Purchases from Agricultural Residents | 53.01 | 79.77 | 160.00 | 189.40 | 226.17 |
| Purchases from Other Production Departments | 2.78 | 4.45 | 8.00 | 9.60 | 10.75 |
| **Purchases by Department** | | | | | |
| Purchases by Commercial Departments | 45.99 | 67.70 | 107.20 | 125.80 | 144.32 |
| *Foreign Trade Departments* | 1.00 | 1.97 | 5.82 | 9.27 | 9.58 |
| Purchases by Industrial and Other Departments | 6.69 | 9.62 | 32.60 | 37.40 | 48.80 |
| Purchases by Non-Agricultural Residents from Agricultural Residents | 3.11 | 6.90 | 28.20 | 35.80 | 43.80 |

## 7.13 *TOTAL VALUE OF IMPORTS AND EXPORTS*

| | Value of Imports and Exports (1,000,000,000 yuan)* | | | Value of Imports and Exports (1,000,000,000 U. S. dollars)* | | |
|---|---|---|---|---|---|---|
| Year | Total | Exports | Imports | Total | Exports | Imports |
| *1978* | 33.51 | 16.77 | 18.74 | 20.64 | 9.75 | 10.89 |
| *1979* | 45.46 | 21.17 | 24.29 | 29.33 | 13.66 | 15.67 |
| *1980* | 57.00 | 27.12 | 29.88 | 38.14 | 18.12 | 20.02 |
| *1981* | 73.53 | 36.76 | 36.77 | 44.02 | 22.01 | 22.01 |
| *1982* | 77.20 | 41.43 | 35.77 | 41.63 | 22.35 | 19.28 |
| *1983* | 86.01 | 43.83 | 42.18 | 43.62 | 22.23 | 21.39 |
| *1984* | 120.11 | 58.06 | 62.05 | 53.55 | 26.14 | 27.41 |
| *1985* | 206.67 | 80.89 | 125.78 | 69.60 | 27.35 | 42.25 |
| *1986* | 258.06 | 108.20 | 149.86 | 73.86 | 30.94 | 42.91 |
| *1987* | 308.69 | 147.18 | 161.51 | 82.73 | 39.49 | 43.24 |

**Note:**
***The conversion of U. S. dollars in RMB is calculated according to the exchange rate stipulated by the Bank of China in each corresponding year. Before 1979, statistics are from foreign trade department; since 1980, customs statistics are used.**

## 7.14 *VOLUME OF PURCHASES OF FARM AND SIDELINE PRODUCTS*

| Commodity | | 1978 | 1980 | 1985 | 1986 | 1987 |
|---|---|---|---|---|---|---|
| *Grain** | 1,000,000 tons | 50.725 | 61.290 | 107.628 | 115.162 | 120.920 |
| *Vegetable Oil** | 1,000,000 tons | 1.105 | 1.910 | 4.098 | 4.270 | 4.411 |
| *Pigs* | 1,000,000 heads | 109.365 | 142.500 | 160.208 | 173.505 | 180.445 |
| *Beef Cattle* | 1,000,000 heads | 1.408 | 2.216 | 4.638 | 5.320 | 6.650 |
| *Mutton Sheep* | 1,000,000 heads | 9.983 | 16.802 | 28.393 | 29.387 | 32.649 |
| *Poultry* | 1,000,000 heads | 162.910 | 266.026 | 671.042 | 747.244 | 821.968 |
| *Fresh Eggs* | 1,000,000 tons | 0.560 | 0.991 | 1.922 | 2.066 | 2.324 |
| *Aquatic Products* | 1,000,000 tons | 2.692 | 2.393 | 3.342 | 3.696 | 4.269 |
| *Tea* | 1,000,000 tons | 0.243 | 0.262 | 0.395 | 0.462 | 0.489 |
| *Honey* | 1,000,000 tons | 0.097 | 0.095 | 0.105 | 0.164 | 0.223 |
| *Cotton* | 1,000,000 tons | 2.096 | 2.610 | 4.319 | 3.794 | 4.071 |
| *Jute, Ambary Hemp* | 1,000,000 tons | 0.533 | 0.544 | 1.203 | 1.105 | 0.829 |
| *Sugarcane* | 1,000,000 tons | 15.579 | 15.840 | 36.875 | 34.663 | 34.975 |
| *Beetroot* | 1,000,000 tons | 2.553 | 5.541 | 8.064 | 7.516 | 7.651 |
| *Flue-Cured Tobacco* | 1,000,000 tons | 1.053 | 0.698 | 1.779 | 1.232 | 1.398 |
| *Silkworm Cocoons* | 1,000,000 tons | 0.171 | 0.243 | 0.361 | 0.356 | 0.350 |
| *Tussah Cocoons* | 1,000,000 tons | 0.055 | 0.072 | 0.024 | 0.022 | 0.036 |
| *Sheep's Wool* | 1,000,000 tons | 0.138 | 0.161 | 0.219 | 0.218 | 0.207 |
| *Goat's Wool* | 1,000,000 tons | 0.008 | 0.008 | 0.005 | 0.007 | 0.005 |
| *Cashmere* | tons | 0.003 | 0.003 | 0.004 | 0.008 | 0.004 |
| *Cattlehide* | 1,000,000 pieces | 5.090 | 4.303 | 5.810 | 5.880 | 6.650 |
| *Sheepskin* | 1,000,000 pieces | 13.950 | 14.696 | 18.020 | 17.890 | 17.031 |
| *Goatskin* | 1,000,000 pieces | 29.030 | 39.010 | 43.260 | 28.300 | 26.489 |
| *Hog Bristles* | 1,000,000 cases | 0.176 | 0.208 | 0.268 | 0.322 | 0.388 |
| *Pork (sausage) Casing* | 1,000,000 units | 83.630 | 107.222 | 91.900 | 151.850 | 123.454 |
| *Tung Oil* | 1,000,000 tons | 0.124 | 0.105 | 0.112 | 0.101 | 0.073 |

**Note:**
*** Grain volume is based on trade (commercial) grain unhusked rice and millet coverted into husked forms according to standard conversion factors.**
**** Vegetable oil includes both oil and oil-bear crops converted into oil according to oil extraction rate.**

## 7.15 NUMBER OF OUTLETS AND EMPLOYEES OF RETAIL SALES, FOOD SERVICE AND SERVICE TRADES

| Item | 1978 | 1980 | 1985 | 1986 | 1987 |
|---|---|---|---|---|---|
| ***Number of Outlets (1,000,000)*** | | | | | |
| *Retail Sales* | 1.048 | 1.463 | 7.783 | 7.967 | 8.814 |
| *Food Service* | 0.117 | 0.299 | 1.353 | 1.420 | 1.551 |
| *Service Trades* | 0.090 | 0.260 | 1.534 | 1.609 | 1.694 |
| ***Number of Employees (1,000,000)*** | | | | | |
| *Retail Sales* | 4.474 | 6.377 | 17.960 | 18.430 | 20.125 |
| *Food Service* | 1.044 | 1.765 | 3.764 | 3.939 | 4.224 |
| *Service Trades* | 0.560 | 1.126 | 3.539 | 3.699 | 3.922 |
| ***Average Number of Persons Served by Each Institution*** | | | | | |
| *Retail Sales* | 914 | 672 | 135 | 133 | 123 |
| *Food Service* | 8189 | 3286 | 776 | 750 | 697 |
| *Service Trades* | 10645 | 3779 | 685 | 663 | 638 |
| ***Average Value of Retail Sales of Each Institution (1,000,000)*** | | | | | |
| *Retail Sales* | 0.130 | 0.121 | 0.042 | 0.047 | 0.049 |
| *Food Service* | 0.047 | 0.027 | 0.015 | 0.016 | 0.018 |
| ***Average Number of Persons Served by Each Staff Member (persons)**** | | | | | |
| *Retail Sales* | 214 | 154 | 58 | 58 | 54 |
| *Food Service* | 918 | 557 | 279 | 270 | 256 |
| *Service Trades* | 1711 | 873 | 297 | 288 | 276 |
| ***Average Value of Retail Sales of Each Staff Member (1,000,000 yuan)*** | | | | | |
| *Retail Sales* | 0.031 | 0.028 | 0.018 | 0.020 | 0.022 |
| *Food Service* | 0.005 | 0.005 | 0.005 | 0.006 | 0.007 |

**Note:**
*** Equal to population divided by number of staff members.**

## 7.16 *VALUE OF IMPORTS AND EXPORTS BY TYPE OF COMMODITY*

(1,000,000,000 U. S. dollars)

| Item | 1985 | | 1986 | | 1987 | |
|---|---|---|---|---|---|---|
| | Exports | Imports | Exports | Imports | Exports | Imports |
| ***Total Value*** | **27.350** | **42.252** | **30.942** | **42.916** | **39.486** | **43.240** |
| ***Primary Goods*** | **13.828** | **5.289** | **11.266** | **5.649** | **13.235** | **6.916** |
| *Food and Living Animals Chiefly for Food* | 3.803 | 1.553 | 4.452 | 1.625 | 4.781 | 2.442 |
| *Beverages and Tobacco* | 0.105 | 0.206 | 0.119 | 0.172 | 0.175 | 0.263 |
| *Non-edible Raw Materials (except fuels)* | 2.653 | 3.236 | 2.908 | 3.143 | 3.651 | 3.321 |
| *Mineral Fuels, Lubricants and Related Materials* | 7.132 | 0.172 | 3.673 | 0.504 | 4.547 | 0.539 |
| *Animal and Vegetable Oils, Fats and Wax* | 0.135 | 0.122 | 0.114 | 0.205 | 0.081 | 0.351 |
| ***Manufactured Goods*** | **13.522** | **36.963** | **19.676** | **37.267** | **26.251** | **36.324** |
| *Chemical and Related Products* | 1.358 | 4.469 | 1.732 | 3.772 | 2.235 | 5.008 |
| *Products of Textile Industries/Rubber and Metal Products* | 4.493 | 11.898 | 5.888 | 11.195 | 8.571 | 9.728 |
| *Machinery and Transportation Equipment* | 0.772 | 16.239 | 1.094 | 16.788 | 1.740 | 14.628 |
| *Miscellaneous Light Industrial Products* | 3.486 | 1.902 | 4.951 | 1.877 | 6.273 | 1.878 |
| *Commodities (not classified elsewhere)* | 3.413 | 2.455 | 6.011 | 3.635 | 7.432 | 5.082 |

## 7.17 *VOLUME AND VALUE OF MAJOR EXPORT COMMODITIES**

| Item | | 1986 | | 1987 | |
|---|---|---|---|---|---|
| | | Volume | Value (1,000,000,000 U. S. dollars) | Volume | Value (1,000,000,000 U. S. dollars) |
| *Live Hogs* | 1,000,000 head | 3.10 | 0.195 | 3.02 | 0.202 |
| *Frozen Pork* | tons | 104670 | 0.175 | 99964 | 0.172 |
| *Aquatic Products* | tons | 166756 | 0.478 | 218210 | 0.703 |
| *Grain* | 1,000,000 tons | 9.42 | 1.312 | 7.08 | 0.983 |
| *Vegetables* | tons | 644038 | 0.274 | 641623 | 0.341 |
| *Fruit* | tons | 223859 | 0.086 | 243792 | 0.101 |
| *Tea* | tons | 172085 | 0.327 | 174274 | 0.362 |
| *Cans* | tons | 445177 | 0.434 | 536981 | 0.536 |
| *Peanuts* | tons | 262419 | 0.157 | 267988 | 0.183 |
| *Raw Silk* | tons | 9394 | 0.233 | 9242 | 0.233 |
| *Cotton* | tons | 558089 | 0.525 | 754576 | 0.756 |
| *Rabbit Hair* | tons | 3556 | 0.130 | 4908 | 0.156 |
| *Tungsten Ore* | tons | 23624 | 0.099 | 23117 | 0.082 |
| *Herb Medicine* | tons | 65018 | 0.170 | 90069 | 0.165 |
| *Coal* | 1,000,000 tons | 9.82 | 0.400 | 13.51 | 0.476 |
| *Petroleum* | 1,000,000 tons | 33.95 | 3.150 | 32.17 | 3.909 |
| *Cotton Yarn* | tons | 228293 | 0.424 | 242957 | 0.535 |
| *Cotton Cloth* | 1,000,000 meters | 2056.27 | 1.270 | 2341.52 | 1.525 |
| *Silk and Satin* | 1,000,000 meters | 132.77 | 0.392 | 150.29 | 0.458 |
| *Household Pottery* | unit | | 0.152 | | 0.193 |
| *Garments* | unit | | 2.420 | | 3.086 |
| *SewingMachines* | unit | 0.60 | 0.017 | 0.59 | 0.020 |
| *Machine Tools* | unit | 48985 | 0.035 | 183866 | 0.079 |
| *Bearings* | 1,000,000 units | | 0.020 | | 0.040 |
| *Electric Fans* | 1,000,000 units | 1.59 | 0.020 | 4.15 | 0.048 |
| *Bicycles* | unit | 0.41 | 0.016 | 0.82 | 0.033 |

**Note:**
*** Data are from customs statistics.**

# 7.18 *VOLUME AND VALUE OF MAJOR IMPORT COMMODITIES**

| Item | | 1986 | | 1987 | |
|---|---|---|---|---|---|
| | | Volume | Value (1,000,000,000 U.S. dollars) | Volume | Value (1,000,000,000 U. S. dollars) |
| *Grain* | 1,000,000 tons | 7.73 | 1.082 | 16.28 | 1.755 |
| *Sugar* | 1,000,000 tons | 1.18 | 0.211 | 1.83 | 0.297 |
| *Vegetable Oil* | tons | 470134 | 0.181 | 849712 | 0.316 |
| *Natural and Synthetic Rubber* | tons | 294780 | 0.258 | 396730 | 0.378 |
| *Lumber* | 1,000,000 cubic meters | 7.15 | 0.606 | 5.62 | 0.421 |
| *Synthetic Fibers Suitable for Spinning* | tons | 300719 | 0.421 | 240635 | 0.370 |
| *Synthetic Fibers,Continuous Filament and Yarn* | tons | 199957 | 0.454 | 85790 | 0.217 |
| *Regenerated Fiber Suitable for Spinning* | tons | 80161 | 0.125 | 112247 | 0.191 |
| *Regenerated Fibers, Continuous Filament and Yarn* | tons | 19173 | 0.077 | 21842 | 0.093 |
| *Sheep Wool* | tons | 152234 | 0.488 | 152504 | 0.543 |
| *Iron Ore Sand* | 1,000,000 tons | 13.72 | 0.334 | 10.87 | 0.269 |
| *Caustic Soda* | 1,000,000 tons | 1.22 | 0.192 | 0.85 | 0.138 |
| *Chemical Fertilizers* | 1,000,000 tons | 5.10 | 0.712 | 10.90 | 1.400 |
| *Chemical Pesticides* | tons | 7495 | 0.041 | 10062 | 0.056 |
| *Paper and Paper Plates* | 1,000,000 meters | | 0.537 | | 0.702 |
| *Steel* | 1,000,000 meters | 18.37 | 5.938 | 12.40 | 4.464 |
| *Copper and Alloys* | tons | 171106 | 0.245 | 75490 | 0.112 |
| *Aluminium and Alloys* | tons | 266241 | 0.331 | 148130 | 0.230 |
| *Zinc and Alloys* | tons | 116945 | 0.091 | 68153 | 0.056 |
| *Machine Tools* | unit | 10079 | 0.362 | 25817 | 0.454 |
| *Ships* | unit | 682 | 0.278 | 955 | 0.149 |
| *Automobiles and Chassises* | unit | 150052 | 1.678 | 90220 | 0.969 |
| *Television Sets* | 1,000,000 | 1.39 | 0.305 | 1.09 | 0.187 |
| *Tape Recorders* | unit | 522527 | 0.034 | 2246395 | 0.049 |
| *Washing Machines* | unit | 184139 | 0.015 | 59908 | 0.005 |
| *Refrigerators* | unit | 483953 | 0.077 | 445116 | 0.080 |

**Note:**
*** Data are from customs statistics.**

# 7.19 *ANNUAL AVERAGE EXCHANGE RATE OF FOREIGN CURRENCIES IN YUAN*

| Foreign Currency | Unit of Foreign Currency | 1986 | | 1987 | |
|---|---|---|---|---|---|
| | | Buying Price (yuan) | Selling Price (yuan) | Buying Price (yuan) | Selling Price (yuan) |
| *Australian Dollar* | 1,000 | 2302.3 | 2313.9 | 2601.3 | 2614.3 |
| *Austrian Schilling* | 1,000 | 227.1 | 228.3 | 293.9 | 295.3 |
| *Canadian Dollar* | 1,000 | 2475.9 | 2488.3 | 2795.9 | 2809.9 |
| *Swiss Franc* | 1,000 | 1933.9 | 1943.5 | 2492.5 | 2504.9 |
| *GFR Mark* | 1,000 | 1614.4 | 1622.4 | 2067.2 | 2077.6 |
| *Danish Krone* | 1,000 | 428.2 | 430.4 | 543.7 | 546.5 |
| *Finnish Markka* | 1,000 | | | 830.4 | 834.6 |
| *French Franc* | 1,000 | 498.9 | 501.5 | 618.0 | 621.0 |
| *Dutch Guilder* | 1,000 | 1416.7 | 1423.9 | 1834.8 | 1844.0 |
| *Norwegian Krone* | 1,000 | 465.7 | 468.1 | 551.3 | 554.1 |
| *Singapore Dollar* | 1,000 | 1584.0 | 1592.0 | 1759.2 | 1768.0 |
| *Swedish Korona* | 1,000 | 484.3 | 486.7 | 585.3 | 588.3 |
| *British Pound* | 1,000 | 5051.3 | 5076.7 | 6083.3 | 6113.7 |
| *US Dollar* | 1,000 | 3444.2 | 3461.4 | 3712.8 | 3731.4 |
| *Hong Kong Dollar* | 1,000 | 441.1 | 443.3 | 476.2 | 478.6 |
| *Italian Lira* | 1,000,000 | 2323 | 2340 | 2870 | 2880 |
| *Japanese Yen* | 1,000,000 | 20641.9 | 20745.3 | 2573.49 | 25863.9 |
| *Belgian Franc* | 1,000,000 | 77540 | 77930 | 99510 | 100010 |

# 7.20 UTILIZATION OF FOREIGN CAPITAL

| Year | Total: Number of Projects | Total: Value (1,000,000,000 U. S. dollars) | Loans: Number of Projects | Loans: Value (1,000,000,000 U. S. dollars) | Direct Investment and Others: Number of Projects | Direct Investment and Others: Value (1,000,000,000 U. S. dollars) |
|---|---|---|---|---|---|---|
| ***Funds under Contracts*** | | | | | | |
| *1979-1987* | **11720** | **60.018** | **298** | **35.709** | **11422** | **24.309** |
| *1979-1982* | 1841 | 18.057 | 27 | 12.522 | 1814 | 5.535 |
| *1983* | 690 | 3.430 | 52 | 1.513 | 638 | 1.917 |
| *1984* | 2204 | 4.791 | 38 | 1.916 | 2166 | 2.875 |
| *1985* | 3145 | 9.867 | 72 | 3.534 | 3073 | 6.333 |
| *1986* | 1551 | 11.737 | 53 | 8.407 | 1498 | 3.330 |
| *1987* | 2239 | 12.136 | 56 | 7.807 | 2233 | 4.319 |
| ***Foreign Funds Already Invested*** | | | | | | |
| *1979-1987* | | **37.686** | | **26.731** | | **10.955** |
| *1979-1982* | | 12.643 | | 10.873 | | 1.770 |
| *1983* | | 1.981 | | 1.065 | | 0.916 |
| *1984* | | 2.705 | | 1.286 | | 1.419 |
| *1985* | | 4.674 | | 2.688 | | 1.959 |
| *1986* | | 7.258 | | 5.014 | | 2.244 |
| *1987* | | 8.452 | | 5.805 | | 2.647 |

# 7.21 FOREIGN INVESTMENT UNDER CONTRACT

| Form of Investments | 1986: Number of Projects | 1986: Value (1,000,000,000 U. S. dollars) | 1987: Number of Projects | 1987: Value (1,000,000,000 U. S. dollars) |
|---|---|---|---|---|
| ***Total*** | **1551** | **11.737** | **2239** | **12.136** |
| | | | | 7.817 |
| ***Loans*** | 53 | 8.407 | 56 | 2.019 |
| *Loans from Foreign Governments* | 42 | 1.444 | 44 | 1.398 |
| *Loans from International Financial Organizations* | 11 | 1.825 | 12 | 4.400 |
| *Others* | | 5.138 | | |
| ***Direct Investment and Others*** | **1498** | **3.330** | **2233** | **4.319** |
| *Joint Ventures* | 892 | 1.375 | 1395 | 1.950 |
| *Contractual Joint Ventures* | 582 | 1.358 | 789 | 1.283 |
| *Foreign Capital Enterprises* | 18 | 0.020 | 46 | 0.471 |
| *Ocean Petroleum Co-development* | 6 | 0.081 | 3 | 0.0004 |
| *Compensation Trade* | | 0.313 | | 0.428 |
| *Others* | | 0.183 | | 0.183 |

## 7.22 NUMBER OF TOURISTS AND FOREIGN EXCHANGE REVENUE FROM TOURISM

| Item | 1978 | 1980 | 1985 | 1986 | 1987 |
|---|---|---|---|---|---|
| ***Total Number of Tourists (1,000,000,000 persons)*** | **1.8092** | **5.7025** | **17.8331** | **22.8194** | **26.9023** |
| *Foreign* | 0.2296 | 0.5291 | 1.3705 | 1.4823 | 1.7278 |
| *Overseas Chinese* | 0.0181 | 0.0344 | 0.0848 | 0.0681 | 0.0871 |
| *Chinese from Hong Kong, Macao and Taiwan Province* | 1.5615 | 5.1390 | 16.3778 | 21.2690 | 25.0874 |
| *Received by International Tourist Agency* | 0.1246 | 0.2187 | 0.4688 | 0.4353 | 0.5050 |
| *Received by China Tourist Agency* | 0.5618 | 0.6947 | 0.7733 | 0.5953 | 0.8477 |
| ***Total Foreign Exchange Revenue from Tourism (1,000,000,000 U. S. Dollars)*** | **0.263** | **0.617** | **1.250** | **1.530** | **1.844** |

## 7.23 NUMBER OF FOREIGN TOURISTS

(1,000,000 persons)

| Country | 1980 | 1986 | 1987 |
|---|---|---|---|
| ***Total*** | **0.5291** | **1.4823** | **1.7278** |
| *Japan* | 0.1693 | 0.4835 | 0.5777 |
| *United States* | 0.1015 | 0.2918 | 0.3153 |
| *Australia* | 0.0284 | 0.0732 | 0.0586 |
| *Britain* | 0.0286 | 0.0794 | 0.0837 |
| *France* | 0.0155 | 0.0403 | 0.0537 |
| *German Federal Republic* | 0.0147 | 0.0482 | 0.0601 |
| *Thailand* | 0.0126 | 0.0398 | 0.0585 |
| *Canada* | 0.0104 | 0.0410 | 0.0523 |
| *Italy* | 0.0068 | 0.0160 | 0.0298 |
| *Democratic People's Republic of Korea* | 0.0079 | 0.0254 | 0.0234 |
| *Switzerland* | 0.0045 | 0.0128 | 0.0170 |
| *Soviet Union* | 0.0045 | 0.0192 | 0.0247 |
| *New Zealand* | 0.0027 | 0.0074 | 0.0109 |

# 7.24 *PRICE INDEXES 1987*

| Item | 1978 = 100 | 1980 = 100 | 1985 = 100 | 1986 = 100 |
|---|---|---|---|---|
| **Cost of Living Index of Staff and Workers** | **156.2** | **142.7** | **116.4** | **108.8** |
| Consumer Goods Price Index | 158.0 | 143.6 | 116.7 | 109.1 |
| Service Price Index | 137.9 | 136.8 | 113.5 | 106.0 |
| **Overall Retail Price Index** | **145.7** | **134.8** | **113.7** | **107.3** |
| Consumer Goods | 147.4 | 134.9 | 114.4 | 107.4 |
| *Food* | 178.5 | 153.0 | 118.2 | 110.1 |
| *Clothes* | 103.3 | 103.8 | 106.8 | 103.5 |
| *Daily Articles* | 120.0 | 117.6 | 112.6 | 106.1 |
| *Cultural Articles* | 95.2 | 98.0 | 103.5 | 102.5 |
| *Books and Newspapers* | | | 115.5 | 101.4 |
| *Medicine* | 125.7 | 122.6 | 106.7 | 104.6 |
| *Fuels* | 118.2 | 117.2 | 107.6 | 103.6 |
| Means of Agricultural Production | 133.8 | 131.7 | 108.2 | 107.0 |
| **Consumer Goods Free Market Index** | **161.8** | **167.1** | **125.7** | **116.3** |
| Grain | 107.9 | 127.4 | 142.4 | 118.1 |
| Edible Vegetable Oil | 99.6 | 108.7 | 120.2 | 110.5 |
| Fresh Vegetables | 198.3 | 193.0 | 127.5 | 119.4 |
| Dried Vegetables | 147.3 | 150.8 | 133.6 | 113.4 |
| Meat, Poultry and Eggs | 167.4 | 175.4 | 127.3 | 119.4 |
| Aquatic Products | 274.6 | 246.7 | 129.3 | 120.8 |
| Fresh Fruits | 211.8 | 204.9 | 121.8 | 108.9 |
| Dried Fruits | | | 125.0 | 118.7 |
| Miscellaneous Goods for Daily Use | 172.8 | 180.8 | 131.3 | 113.7 |
| Firewood and Grass | 207.4 | 197.1 | 135.1 | 116.4 |
| **Overall Purchase Price Index of Farm and Sideline Products** | **198.8** | **152.1** | **119.2** | **112.0** |
| Grain | 239.4 | 170.0 | 118.7 | 108.0 |
| Industry Crops | 163.9 | 119.8 | 107.0 | 103.3 |
| Livestock Products | 205.1 | 160.7 | 121.4 | 117.9 |
| Other Farm and Sideline Products | 205.0 | 176.9 | 125.1 | 116.2 |

## 7.25 *OVERALL RETAIL PRICE INDEX AND COST OF LIVING INDEX OF STAFF AND WORKERS BY REGION 1987*

| | Overall Retail Price Index | | | Cost of Living Index of Staff and Workers | | |
|---|---|---|---|---|---|---|
| Region | 1978 =100 | 1980 =100 | 1986 = 100 | 1978 = 100 | 1980 =100 | 1986 = 100 |
| ***National Average*** | **145.7** | **134.8** | **107.3** | **156.2** | **142.7** | **108.8** |
| *Beijing* | 158.6 | 146.0 | 108.7 | 155.8 | 144.4 | 108.6 |
| *Tianjin* | 145.3 | 136.2 | 106.9 | 142.5 | 134.2 | 106.8 |
| *Hebei* | 141.1 | 132.2 | 108.3 | 149.1 | 138.9 | 108.2 |
| *Shanxi* | 138.1 | 132.8 | 107.5 | 148.0 | 138.9 | 108.5 |
| *Inner Mongolia* | 144.6 | 134.5 | 108.1 | 148.9 | 137.1 | 108.5 |
| *Liaoning* | 147.7 | 137.8 | 109.0 | 157.1 | 142.6 | 109.8 |
| *Jilin* | 149.9 | 139.2 | 107.5 | 157.5 | 144.1 | 107.6 |
| *Heilongjiang* | 156.1 | 145.2 | 109.6 | 161.1 | 146.4 | 109.7 |
| *Shanghai* | 151.3 | 140.7 | 108.8 | 147.3 | 137.8 | 108.1 |
| *Jiangsu* | 144.8 | 135.5 | 109.3 | 151.1 | 137.4 | 110.5 |
| *Zhejiang* | 157.8 | 142.8 | 109.5 | 168.0 | 149.8 | 110.9 |
| *Anhui* | 137.1 | 130.1 | 109.7 | 143.0 | 134.3 | 109.9 |
| *Fujian* | 153.4 | 142.1 | 109.4 | 165.5 | 151.4 | 110.6 |
| *Jiangxi* | 144.0 | 136.9 | 106.9 | 150.1 | 139.1 | 107.9 |
| *Shandong* | 133.1 | 127.7 | 108.0 | 140.7 | 133.2 | 109.1 |
| *Henan* | 134.3 | 128.3 | 108.1 | 142.9 | 134.3 | 107.8 |
| *Hubei* | 136.4 | 128.5 | 107.6 | 150.1 | 137.0 | 108.7 |
| *Hunan* | 158.9 | 140.6 | 110.6 | 170.6 | 145.4 | 111.3 |
| *Guangdong* | 169.2 | 151.6 | 111.7 | 180.9 | 157.9 | 112.8 |
| *Guangxi* | 158.4 | 141.8 | 108.0 | 179.1 | 154.6 | 110.2 |
| *Sichuan* | 145.8 | 127.9 | 107.5 | 158.5 | 137.2 | 110.1 |
| *Guizhou* | 144.7 | 131.2 | 107.3 | 158.1 | 141.5 | 109.7 |
| *Yunnan* | 137.7 | 119.7 | 106.6 | 147.1 | 134.9 | 107.4 |
| *Tibet* | | | | | | |
| *Shaanxi* | 142.3 | 133.5 | 108.6 | 147.9 | 137.7 | 109.2 |
| *Gansu* | 140.2 | 132.2 | 107.4 | 153.7 | 140.7 | 108.4 |
| *Qinghai* | 143.9 | 135.4 | 107.3 | 146.6 | 135.9 | 107.8 |
| *Ningxia* | 142.8 | 133.3 | 108.0 | 153.0 | 139.5 | 109.9 |
| *Xinjiang* | 141.9 | 131.7 | 107.1 | 145.8 | 136.2 | 108.5 |

*Section VIII*

# PEOPLE'S LIVELIHOOD

# 8.1 PERSONAL INCOME AND QUALITY OF LIFE

| Item | | 1978 | 1980 | 1986 | 1987 |
|---|---|---|---|---|---|
| ***Income of Rural and Urban Residents*** | | | | | |
| *Average per Capita Annual Income of Farmers* | yuan | 134.00 | 191.00 | 424.00 | 463.00 |
| *Average per Capita Annual Wage of Staff and Workers* | yuan | 614.00 | 762.00 | 1329.00 | 1459.00 |
| *Average per Capita Annual Income for Living Expenses of Urban Households* | yuan | 316 | 439 | 828 | 916 |
| ***Average per Capita Living Space*** | | | | | |
| *Urban* | square meters | 4.2 | 5 | 8 | 8.6 |
| *Rural* | square meters | 8.1 | 9.4 | 15.3 | 16 |
| ***Savings Deposits*** | | | | | |
| *Total Bank Savings Balances of Rural and Urban Residents at Year End* | 1,000,000,000 yuan | 21.06 | 39.95 | 223.79 | 307.53 |
| *Average per Capita Bank Savings* | yuan | 21.9 | 40.5 | 210 | 284.6 |
| ***Culture*** | | | | | |
| *Number of Televisions Sets per 1000 Persons* | unit | 3 | 9 | 87 | 107 |
| *Number of Radios per 1000 Persons* | unit | 78 | 121 | 240 | 2.1 |
| *Number of Daily Newspapers per 1000 Persons* | copy | 32 | 39 | 50 | 53 |
| *Per Capita Annual Books or Magazines* | copy | 4.8 | 5.8 | 7.2 | 8.3 |
| ***Public Health*** | | | | | |
| *Hospital Beds per 1,000 Persons* | unit | 1.94 | 2.02 | 2.16 | 2.23 |
| *Doctors per 1,000 Persons* | person | 1.08 | 1.17 | 1.36 | 1.38 |
| ***Employment*** | | | | | |
| *Average Number of Dependents per Urban Employee* | person | 2.06 | 1.77 | 1.8 | 1.79 |
| ***Retail Sales, Food Service and Service Trades**** | | | | | |
| *Number of Trades and Service units per 1,000,000 Persons* | unit | 1300 | 2050 | 10320 | 11160 |
| *Number of Service Personnel per 1,000,000 Persons* | person | 6300 | 9390 | 24470 | 26160 |

**Note:**
*** Number of units includes individual owned units; number of personnel includes individual laborers.**

## 8.2 *AVERAGE ANNUAL PERSONAL CONSUMPTION **

| Year | Per Capita Consumption (yuan) (at current prices) | | | Index (at comparable prices) | | |
|---|---|---|---|---|---|---|
| | National Average | Rural Areas | Urban Areas** | National Average | Rural Areas | Urban Areas** |
| *1978* | 175 | 132 | 383 | 100 | 100.0 | 100.0 |
| *1979* | 197 | 152 | 406 | 109.4 | 106.9 | 111.6 |
| *1980* | 227 | 173 | 468 | 119.8 | 117.3 | 119.6 |
| *1981* | 249 | 194 | 487 | 128.0 | 128.3 | 121.4 |
| *1982* | 267 | 212 | 500 | 134.6 | 137.7 | 122.3 |
| *1983* | 289 | 234 | 531 | 144.0 | 149.8 | 127.1 |
| *1984* | 329 | 266 | 599 | 160.1 | 166.6 | 139.7 |
| *1985* | 405 | 323 | 747 | 181.4 | 189.8 | 155.6 |
| *1986* | 450 | 352 | 851 | 189.1 | 194.4 | 165.8 |
| *1987* | 504 | 386 | 978 | 197.1 | 200.4 | 175.1 |

**Note:**
*** Figures in this table are based on personal consumption in national income and annual average population figures.**

**** Urban area figures include persons living in rural areas and not engaging in agricultural activities and not working in village-run industries or other service units.**

## 8.3 *AVERAGE ANNUAL GROWTH RATE OF PERSONAL CONSUMPTION*

| Year | National Average | Rural Areas | Urban Areas** |
|---|---|---|---|
| *1976-1980* | 4.8 | 4.1 | 4.9 |
| *1981-1985* | 8.7 | 10.1 | 5.4 |
| *1979-1987* | 7.8 | 8.0 | 6.4 |

## 8.4 *PER CAPITA CONSUMPTION OF CONSUMER GOODS **

| Item | | 1978 | 1980 | 1985 | 1986 | 1987 |
|---|---|---|---|---|---|---|
| *Grain* | kilograms | 195.5 | 213.8 | 253.2 | 254.8 | 251.4 |
| *Vegetable Oil* | kilograms | 1.6 | 2.3 | 5.1 | 5.2 | 5.4 |
| *Pork* | kilograms | 7.7 | 11.2 | 13.9 | 14.3 | 14.5 |
| *Fresh Eggs* | kilograms | 2.0 | 2.3 | 5.0 | 5.2 | 5.6 |
| *Sugar* | kilograms | 3.4 | 3.8 | 5.6 | 6.1 | 6.7 |
| *Cloth* | meters | 8.0 | 10.0 | 11.6 | 11.1 | 11.3 |

**Note:**
*** The volume of consumer goods include those bought from markets and those produced for self-consumption by rural laborers.**

## 8.5 *OWNERSHIP OF DURABLE CONSUMER GOODS*
(1,000,000 units) (year end figures)

| Item | 1978 | 1980 | 1985 | 1986 | 1987 |
|---|---|---|---|---|---|
| ***Selected Durable Goods*** | | | | | |
| *Sewing Machines* | 33.96 | 46.04 | 98.33 | 109.36 | 119.14 |
| *Wrist Watches* | 82.06 | 127.65 | 360.69 | 413.73 | 462.84 |
| *Bicycles* | 74.26 | 96.17 | 223.64 | 258.03 | 293.13 |
| *Radios* | 75.46 | 119.10 | 241.81 | 253.90 | 260.67 |
| *Television Sets* | 3.04 | 9.02 | 69.65 | 92.14 | 116.00 |
| ***Average per 100 Population*** | | | | | |
| *Sewing Machines* | 3.5 | 4.7 | 9.4 | 10.3 | 11.0 |
| *Wrist Watches* | 8.5 | 12.9 | 34.5 | 39.0 | 42.8 |
| *Bicycles* | 7.7 | 9.7 | 21.4 | 24.3 | 27.1 |
| *Radios* | 7.8 | 12.1 | 23.1 | 24.0 | 24.1 |
| *Television Sets* | 0.3 | 0.9 | 6.7 | 8.7 | 10.7 |

## 8.6 *WAGE BILL OF STAFF AND WORKERS*

(1,000,000,000 yuan)

| Item | 1978 | 1980 | 1985 | 1986 | 1987 |
|---|---|---|---|---|---|
| **Total Average Wages** | **56.88** | **77.25** | **138.30** | **165.97** | **188.13** |
| **State Owned Enterprises** | 46.86 | 62.79 | 106.48 | 128.85 | 145.86 |
| **Collective Owned Enterprises** | 10.02 | 14.46 | 31.23 | 36.28 | 40.90 |
| **Other Units** | | | 0.59 | 0.84 | 1.27 |
| Bonuses and Overfulfilled Piecework Quota Wage | 1.13 | 7.03 | 20.86 | 24.25 | 31.57 |
| *State owned Enterprises* | 1.13 | 6.09 | 15.47 | 18.92 | 24.64 |
| *Collective Owned Enterprises* | | 0.94 | 5.26 | 5.16 | 6.65 |
| *Other Units* | | | 0.13 | 0.17 | 0.28 |
| Per Capita Bonuses and Overfulfilled Piecework Quota Wages (yuan) | 12 | 69 | 173 | 194 | 245 |
| **Average Annual Wages (yuan)** | **615** | **762** | **1148** | **1329** | **1459** |
| State Owned Enterprises | 644 | 803 | 1213 | 1414 | 1546 |
| Collective Owned Enterprises | 506 | 624 | 968 | 10920 | 1207 |
| Other Units | | | 1436 | 16290 | 1879 |

**Note:**
**1985-1987 total wages include subsidies for meat price increase, etc..**

## 8.7 *PERCENTAGE INCREASE IN TOTAL WAGES*

| Item | 1987 over 1980 | | 1987 over 1986 | |
|---|---|---|---|---|
| | Increment (1,000,000,000 yuan) | Percentage | Increment (1,000,000,000 yuan) | Percentage |
| ***Total Wage Increase*** | **110.8** | **100.0** | **22.1** | **100.0** |
| *Standard Salary Increases* | 61.8 | 55.8 | 10.3 | 46.6 |
| *Bonuses and Overfulfilled Piecework Quota Wages* | 24.5 | 22.1 | 7.3 | 33.0 |
| *Subsidies Increase* | 22.4 | 20.2 | 3.8 | 17.2 |
| *Overtime Payments Increase* | 2.6 | 2.3 | 0.7 | 3.2 |
| *Others* | -0.5 | 0.4 | | |

## 8.8 PERCENTAGE OF TOTAL WAGES OF STAFF AND WORKERS OF STATE OWNED UNITS

| Item | 1978 | 1980 | 1985 | 1986 | 1987 |
|---|---|---|---|---|---|
| **Total Wages** | **100.0** | **100.0** | **100.0** | **100.0** | **100.0** |
| Time Wages | 85.0 | 69.8 | 59.5 | 58.7 | 56.5 |
| Piece Wages | 0.8 | 3.2 | 9.9 | 9.1 | 9.6 |
| *Overfulfilment Wages** | 0.1 | 0.6 | 2.2 | 2.0 | 2.2 |
| Bonuses | 2.3 | 9.1 | 12.9 | 13.4 | 15.4 |
| Subsidies | 6.5 | 14.1 | 15.2 | 15.2 | 15.5 |
| Overtime Payments | 2.0 | 1.6 | 1.6 | 1.9 | 2.0 |
| Other | 3.4 | 2.2 | 0.9 | 1.7 | 1.0 |

**Note:**
*** This refers to overfulfilment wages of piece work.**

## 8.9 AVERAGE WAGE OF STAFF AND WORKERS OF STATE OWNED UNITS

| Item | Average Annual Wage (yuan) | Index (1978 = 100) | |
|---|---|---|---|
| | | Nominal Wage | Real Wage* |
| *1978* | 644 | 100.0 | 100.0 |
| *1979* | 705 | 109.5 | 107.6 |
| *1980* | 803 | 124.7 | 114.1 |
| *1981* | 812 | 126.1 | 112.6 |
| *1982* | 830 | 129.8 | 113.5 |
| *1983* | 865 | 134.3 | 115.1 |
| *1984* | 1034 | 160.6 | 136.5 |
| *1985* | 1213 | 188.4 | 140.4 |
| *1986* | 1414 | 219.6 | 153.0 |
| *1987* | 1546 | 240.1 | 153.7 |

**Note:**
*** Real wage index is calculated by deflating nominal wage index using cost of living index of staff and workers.**

# 8.10 *BASIC INDICATORS OF URBAN HOUSEHOLDS FROM SAMPLE SURVEY*

| Item | | **1981 | 1985 | 1986 | 1987 |
|---|---|---|---|---|---|
| ***Number of Households Surveyed*** | household | 8715.00 | 24338.00 | 27024.00 | 32855.00 |
| ***Average Number of Persons per Household*** | person | 4.24 | 3.89 | 3.82 | 3.74 |
| ***Average Number of Employees per Household*** | person | 2.39 | 2.15 | 2.12 | 2.09 |
| ***Percentage of Employees per Household*** | person | 56.37 | 55.27 | 55.50 | 55.88 |
| ***Average Number of Dependents per Employee*** | person | 1.77 | 1.71 | 1.80 | 1.79 |
| ***Annual Per Capita Income*** | yuan | 500.40 | 748.92 | 909.96 | 1012.20 |
| *Per Capita Disposable Income** | yuan | 458.04 | 685.32 | 827.88 | 915.96 |
| ***Distribution of Total Household Expenditures*** | | | | | |
| *Lowest Income Households (10%)* | % | ***15.56 | 6.52 | 6.61 | 6.64 |
| *Second Lowest Income Households (10%)* | % | | 7.85 | 8.41 | 7.93 |
| *Middle Lower Income Households (20%)* | % | 17.84 | 17.63 | 17.76 | 17.87 |
| *Middle Income Households (20%)* | % | 19.04 | 19.45 | 19.43 | 19.79 |
| *Upper Middle Income Households (20%)* | % | 21.25 | 21.70 | 21.62 | 21.26 |
| *Upper Income Households (10%)* | % | ****26.31 | 12.32 | 12.04 | 12.05 |
| *Top Income Households (10%)* | % | | 14.53 | 14.13 | 14.49 |
| ***Average Annual Living Expenditure* Per Capita*** | yuan | 456.84 | 673.20 | 798.96 | 884.40 |

**Note:**

*** Disposable income refers to the actual income of the surveyed household available for daily life, excluding financial support and gifts to others, and the allowance received for help in compiling the survey.**

**** The figures in 1981 only refer to cities. From 1985, the statistics include a sample survey of households in cities and countytowns.**

***** This number covers two groups: the lowest income households and the second lowest income households.**

****** This number covers two groups: the upper income households and the top income households.**

# 8.11 *AVERAGE ANNUAL PER CAPITA LIVING EXPENDITURE OF URBAN HOUSEHOLDS*

| Item | 1981 (yuan) | 1986 (yuan) | 1987 (yuan) | 1981 (percentage) | 1986 (percentage) | 1987 (percentage) |
|---|---|---|---|---|---|---|
| **Total Living Expenditure** | **456.84** | **798.96** | **884.40** | **100.00** | **100.00** | **100.00** |
| **Purchase of Commodities** | **420.36** | **734.64** | **809.27** | **92.01** | **91.95** | **91.50** |
| Food | 258.84 | 418.92 | 472.93 | 56.66 | 52.43 | 53.47 |
| *Grain* | 59.16 | 64.80 | 66.92 | 12.95 | 8.11 | 7.57 |
| *Foodstuffs* | 140.28 | 242.52 | 276.86 | 30.17 | 30.35 | 31.30 |
| *Tobacco, Liquor, Tea* | 23.16 | 44.52 | 50.72 | 5.07 | 5.57 | 5.73 |
| *Other* | 36.24 | 67.08 | 78.43 | 7.93 | 8.40 | 8.87 |
| Clothing | 67.56 | 113.04 | 121.09 | 14.79 | 14.15 | 13.69 |
| Articles for Daily Use | 43.68 | 88.92 | 100.57 | 9.56 | 11.13 | 11.37 |
| Articles for Culture and Recreation | 25.92 | 54.48 | 49.58 | 5.67 | 6.82 | 5.61 |
| Books, Newspapers and Magazines | 4.32 | 6.96 | 7.48 | 0.95 | 0.87 | 0.85 |
| Medicine and Medical Products | 2.76 | 7.56 | 8.87 | 0.60 | 0.95 | 1.00 |
| Housing and Construction Materials | 0.72 | 18.96 | 22.56 | 0.16 | 2.37 | 2.55 |
| Fuel | 8.88 | 12.00 | 12.27 | 1.94 | 1.50 | 1.39 |
| Other | 7.68 | 13.80 | 13.92 | 1.68 | 1.73 | 1.57 |
| **Non-Commodity Expenditure** | **36.48** | **64.32** | **75.13** | **7.99** | **8.05** | **8.50** |
| Rent | 6.36 | 7.20 | 7.74 | 1.39 | 0.90 | 0.88 |
| Water and Electricity | 4.44 | 8.76 | 10.32 | 0.97 | 1.10 | 1.17 |
| Gas | 0.84 | 0.96 | 1.35 | 0.18 | 0.12 | 0.15 |
| Education | 2.64 | 8.76 | 11.28 | 0.58 | 1.10 | 1.28 |
| Child Care | 2.76 | 3.72 | 3.70 | 0.60 | 0.47 | 0.42 |
| Transportation | 6.12 | 8.40 | 9.10 | 1.34 | 1.05 | 1.03 |
| Posts and Telecommunications | 0.48 | 0.72 | 0.82 | 0.11 | 0.09 | 0.09 |
| Culture and Recreation | 2.88 | 2.88 | 3.01 | 0.63 | 0.36 | 0.34 |
| Repairs and Services | 5.16 | 8.88 | 15.28 | 1.13 | 1.11 | 1.73 |
| Medical Treatment | 1.44 | 2.16 | 2.56 | 0.32 | 0.27 | 0.29 |
| Other | 3.36 | 11.88 | 9.97 | 0.74 | 1.48 | 1.12 |

## 8.12 *AVERAGE POSSESSION OF DURABLE CONSUMER GOODS PER ONE HUNDRED URBAN HOUSEHOLDS*

(year end figures) (unit)

| Item | 1981 | 1985 | 1986 | 1987 |
|---|---|---|---|---|
| *Bicycles* | 135.90 | 152.27 | 163.45 | 176.53 |
| *Sewing Machines* | 70.41 | 70.82 | 73.85 | 74.88 |
| *Wrist Watches* | 240.76 | 274.76 | 298.96 | 314.91 |
| *Electric Fans* | 42.62 | 73.91 | 90.01 | 103.92 |
| *Washing Machines* | 6.34 | 48.29 | 59.70 | 66.77 |
| *Refrigerators* | 0.22 | 6.58 | 12.71 | 19.91 |
| *Wardrobes* | 86.09 | 102.08 | 104.03 | 105.03 |
| *Sofas* | 89.33 | 131.49 | 140.00 | 145.08 |
| *Desks* | 55.17 | 80.06 | 84.10 | 87.18 |
| *Radios* | 100.52 | 74.52 | 68.71 | 67.01 |
| *Black/White Television Sets* | 57.06 | 66.86 | 65.42 | 64.77 |
| *Color Television Sets* | 0.59 | 17.21 | 27.41 | 34.63 |
| *Recorders* | 12.97 | 41.16 | 51.67 | 57.38 |
| *Cameras* | 4.29 | 8.52 | 11.91 | 14.34 |

## 8.13 *BASIC INDICATORS OF RURAL HOUSEHOLDS FROM SAMPLE SURVEY*

| Item | | 1978 | 1980 | 1985 | 1986 | 1987 |
|---|---|---|---|---|---|---|
| ***Number of Households Surveyed*** | household | 6095 | 15914 | 66642 | 66836 | 66912 |
| ***Permanent Residents in Households Surveyed*** | person | 34961 | 88090 | 341525 | 338736 | 334970 |
| ***Permanent Residents per Households*** | person | 5.74 | 5.54 | 5.12 | 5.07 | 5.01 |
| ***Average Number of Laborers per Household*** | person | 2.27 | 2.45 | 2.95 | 2.95 | 2.95 |
| ***Average Number of Dependents per Laborer*** | person | 2.53 | 2.26 | 1.74 | 1.72 | 1.70 |
| ***Average Number of Rooms Built per Household in Survey Year*** | room | 0.11 | 0.23 | 0.30 | 0.29 | 0.27 |
| ***Average Number of Rooms per Household at Year End*** | room | 3.64 | 4.06 | 5.11 | 5.27 | 5.36 |
| ***Total Floor Space per Capita at Year End*** | square meters | 10.17 | 11.59 | 17.34 | 18.09 | 18.88 |
| *Living Space* | square meters | 8.10 | 9.40 | 14.70 | 15.29 | 16.00 |

## 8.14 AVERAGE PER CAPITA INCOME OF RURAL HOUSEHOLDS

| Item | 1978 | 1980 | 1985 | 1986 | 1987 |
|---|---|---|---|---|---|
| ***Average Per Capita Annual Income (yuan)*** | **133.57** | **191.33** | **397.60** | **423.76** | **462.55** |
| *Income from Collective-run Business* | 88.53 | 108.37 | 33.37 | 36.15 | 42.09 |
| *Income from Local Economic Associations* | | | 3.69 | 2.92 | 3.49 |
| *Net Income from Household-run Business* | 35.79 | 62.55 | 322.53 | 345.28 | 383.57 |
| *Other Non-credit Income* | 9.25 | 20.41 | 38.01 | 39.41 | 33.40 |
| ***Percentage (total income = 100)*** | | | | | |
| *Income from Collective-run Business* | **66.30** | **56.60** | **8.40** | **8.50** | **9.10** |
| *Income from Local Economic Associations* | | | 0.90 | 0.70 | 0.80 |
| *Net Income from Household-run Business* | 26.80 | 32.70 | 81.10 | 81.50 | 82.90 |
| *Other Non-credit Income* | 6.90 | 10.70 | 9.60 | 9.30 | 7.20 |

## 8.15 AVERAGE PER CAPITA LIVING EXPENDITURE OF RURAL HOUSEHOLDS

| Item | 1978 | 1980 | 1985 | 1986 | 1987 |
|---|---|---|---|---|---|
| ***Total Per Capita Annual Living Expenditure (yuan)*** | **116.06** | **162.21** | **317.42** | **356.95** | **398.29** |
| ***Expenditure on Consumer Goods*** | 112.90 | 157.95 | 308.35 | 345.71 | 378.14 |
| *Food* | 78.59 | 100.19 | 183.33 | 201.17 | 219.67 |
| *Clothing* | 14.74 | 19.99 | 31.34 | 33.74 | 34.23 |
| *Fuel* | 8.28 | 9.66 | 18.16 | 18.57 | 19.29 |
| *Housing* | 3.67 | 12.80 | 39.46 | 51.23 | 57.76 |
| *Daily Articles and Others* | 7.62 | 15.31 | 36.06 | 41.00 | 47.19 |
| ***Cultural Articles and Services*** | 3.16 | 4.26 | 9.07 | 11.24 | 20.15 |
| ***Percentage of Total*** | **100.0** | **100.0** | **100.0** | **100.0** | **100.0** |
| ***Expenditure on Consumer Goods*** | 97.3 | 97.4 | 97.1 | 96.9 | 94.9 |
| *Food* | 67.7 | 61.8 | 57.7 | 56.3 | 55.2 |
| *Clothing* | 12.7 | 12.3 | 9.9 | 9.5 | 8.6 |
| *Fuel* | 7.1 | 6.0 | 5.7 | 5.2 | 4.8 |
| *Housing* | 3.2 | 7.9 | 12.4 | 14.4 | 14.5 |
| *Daily Articles and Others* | 6.6 | 9.4 | 11.4 | 11.5 | 11.8 |
| ***Cultural Articles and Services*** | 2.7 | 2.6 | 2.9 | 3.1 | 5.1 |

**Note:**
*** Living expenditure of rural households includes items purchased from markets and self-produced for consumption.**

## 8.16 *VALUE OF PURCHASE AS PERCENTAGE OF TOTAL LIVING EXPENDITURE OF RURAL HOUSEHOLDS*

(total expenditure for each item = 100)

| Item | 1978 | 1980 | 1985 | 1986 | 1987 |
|---|---|---|---|---|---|
| ***Value of Purchased Items*** | 39.7 | 50.4 | 60.2 | 62.8 | 64.5 |
| *Food* | 24.1 | 31.1 | 41.7 | 44.1 | 46.4 |
| *Clothing* | 89.0 | 98.1 | 98.0 | 98.1 | 98.0 |
| *Fuel* | 31.9 | 28.7 | 21.6 | 21.9 | 23.1 |
| *Housing* | 95.1 | 88.8 | 98.0 | 98.2 | 98.3 |
| *Daily Articles and Others* | 87.7 | 96.3 | 99.5 | 99.6 | 99.7 |

**Note:**
*** Total living expenditure of rural households includes items purchased from markets and self-produced for consumption. Figures in this table show the proportion of the value of purchased goods in total expenditure for each category.**

## 8.17 *AVERAGE PER CAPITA CONSUMPTION OF CONSUMER GOODS OF RURAL HOUSEHOLDS*

| Item | | 1978 | 1980 | 1985 | 1986 | 1987 |
|---|---|---|---|---|---|---|
| ***Grain*** | kilograms | 248.00 | 257.00 | 257.00 | 259.00 | 259.00 |
| *Wheat and Rice* | kilograms | 123.00 | 163.00 | 209.00 | 212.00 | 211.00 |
| ***Vegetables*** | kilograms | 142.00 | 127.00 | 131.00 | 134.00 | 130.00 |
| ***Vegetable Oil*** | kilograms | 1.97 | 2.49 | 4.04 | 4.19 | 4.69 |
| ***Meat*** | kilograms | 5.76 | 7.75 | 10.97 | 11.79 | 11.65 |
| ***Poultry*** | kilograms | 0.25 | 0.66 | 1.03 | 1.14 | 1.15 |
| ***Eggs*** | kilograms | 0.80 | 1.20 | 2.05 | 2.08 | 2.25 |
| ***Fish and Shrimp*** | kilograms | 0.84 | 1.10 | 1.64 | 1.87 | 1.96 |
| ***Sugar*** | kilograms | 0.73 | 1.06 | 1.46 | 1.59 | 1.70 |
| ***Liquor*** | kilograms | 1.22 | 1.89 | 4.37 | 4.96 | 5.48 |
| ***Cotton*** | kilograms | 0.40 | 0.38 | 0.43 | 0.34 | 0.52 |
| ***Cotton Cloth*** | meters | 5.63 | 4.30 | 2.54 | 1.98 | 1.57 |
| ***Chemical Fiber Cloth*** | meters | 0.41 | 0.94 | 2.50 | 2.53 | 2.28 |
| ***Woolen Cloth*** | meters | 0.02 | 0.06 | 0.14 | 0.13 | 0.12 |
| ***Silk and Satin*** | meters | 0.02 | 0.06 | 0.07 | 0.06 | 0.05 |
| ***Knitting Wool and Woolen Clothes*** | kilograms | 0.02 | 0.05 | 0.04 | 0.05 | 0.07 |
| ***Rubber, Gym and Leather Shoes*** | pair | 0.32 | 0.51 | 0.55 | 0.66 | 0.68 |

## 8.18 *OWNERSHIP OF DURABLE CONSUMER GOODS PER ONE HUNDRED RURAL HOUSEHOLDS*

(year-end figures) (unit)

| Item | 1978 | 1980 | 1985 | 1986 | 1987 |
|---|---|---|---|---|---|
| ***Bicycles*** | 30.73 | 36.87 | 80.64 | 90.31 | 98.52 |
| ***Sewing Machines*** | 19.80 | 23.31 | 43.21 | 46.99 | 49.79 |
| ***Radios*** | 17.44 | 33.54 | 54.19 | 54.24 | 52.98 |
| ***Clocks*** | 24.33 | 30.95 | 33.94 | 50.74 | 46.92 |
| ***Wrist Watches*** | 27.42 | 37.58 | 126.32 | 145.06 | 161.22 |
| *Electronic Watches* | | | | 16.13 | 22.56 |
| ***Television Sets*** | | 0.39 | 11.74 | 17.28 | 24.38 |
| *Color TV Sets* | | | | 1.52 | 2.34 |

## 8.19 *TOTAL RURAL AND URBAN PERSONAL SAVINGS DEPOSITS*

(year-end balance) (1,000,000,000 yuan)

| Year | Total | Urban Savings Deposits | | Rural Savings Deposits |
|---|---|---|---|---|
| | | Subtotal | Long Term Deposits | |
| *1978* | 21.06 | 15.49 | 12.89 | 5.57 |
| *1979* | 28.10 | 20.26 | 16.64 | 7.84 |
| *1980* | 39.95 | 28.25 | 22.86 | 11.70 |
| *1981* | 52.37 | 35.41 | 28.94 | 16.96 |
| *1982* | 67.54 | 44.73 | 36.52 | 22.81 |
| *1983* | 89.25 | 57.26 | 46.39 | 31.99 |
| *1984* | 121.47 | 77.66 | 61.53 | 43.81 |
| *1985* | 162.26 | 105.78 | 84.12 | 56.48 |
| *1986* | 223.76 | 147.15 | 118.93 | 76.61 |
| *1987* | 307.53 | 206.74 | 164.79 | 100.79 |

*Section IX*

# EDUCATION, SCIENCE, CULTURE, PUBLIC HEALTH AND SPORTS

## 9.1 *STUDENT ENROLLMENT BY LEVEL*

(1,000,000 persons)

| Year | Total | Institutions of Higher Learning* | Secondary Schools | | | Primary Schools |
|---|---|---|---|---|---|---|
| | | | *Subtotal* | *Specialized Secondary Schools* | *Regular Secondary Schools* | |
| *1978* | 213.468 | 0.856 | 66.372 | 0.889 | 65.483 | 146.240 |
| *1979* | 207.898 | 1.020 | 60.249 | 1.199 | 59.050 | 146.629 |
| *1980* | 204.192 | 1.144 | 56.778 | 1.243 | 55.081 | 146.270 |
| *1981* | 194.753 | 1.279 | 50.146 | 1.069 | 48.596 | 143.328 |
| *1982* | 187.902 | 1.154 | 47.028 | 1.039 | 45.285 | 139.720 |
| *1983* | 183.327 | 1.207 | 46.341 | 1.143 | 43.978 | 135.780 |
| *1984* | 185.576 | 1.396 | 48.609 | 1.322 | 45.542 | 135.571 |
| *1985* | 186.331 | 1.703 | 50.926 | 1.571 | 47.060 | 133.702 |
| *1986* | 186.921 | 1.880 | 53.216 | 1.757 | 48.899 | 131.825 |
| *1987* | 184.355 | 1.959 | 54.037 | 1.874 | 49.481 | 128.359 |

**Note:**
***Include all types of post-secondary level learning institutions.**

## 9.2 *STUDENT ENROLLMENT PER THOUSAND POPULATION BY LEVEL*

| Year | Students as Percentage of the Total Population | College Students per 1,000 Population | Secondary School Students per 1,000 Population | Primary School Students per 1,000 Population | Collge Students as Percentage of Total Students | Secondary School Students as Percentage of Total Students | Primary School Students as Percentage of Total Students |
|---|---|---|---|---|---|---|---|
| *1978* | 22.28 | 0.89 | 69.3 | 152.6 | 0.4 | 31.1 | 68.5 |
| *1979* | 21.41 | 1.05 | 62.1 | 151.0 | 0.5 | 29.0 | 70.5 |
| *1980* | 20.78 | 1.16 | 57.8 | 148.9 | 0.6 | 27.8 | 71.6 |
| *1981* | 19.55 | 1.28 | 50.3 | 143.9 | 0.7 | 25.7 | 73.6 |
| *1982* | 18.57 | 1.14 | 46.5 | 138.1 | 0.6 | 25.0 | 74.4 |
| *1983* | 17.91 | 1.18 | 45.3 | 132.6 | 0.6 | 25.3 | 74.1 |
| *1984* | 17.93 | 1.35 | 47.0 | 131.0 | 0.7 | 26.2 | 73.1 |
| *1985* | 17.81 | 1.63 | 48.7 | 127.8 | 0.9 | 27.3 | 71.8 |
| *1986* | 17.61 | 1.77 | 50.1 | 124.2 | 1.0 | 28.5 | 70.5 |
| *1987* | 17.11 | 1.82 | 50.2 | 119.1 | 1.1 | 29.3 | 69.6 |

## 9.3 *NUMBER OF GRADUATES FROM INSTITUTIONS OF HIGHER LEARNING AND SPECIALIZED SECONDARY SCHOOLS*

(1,000 persons)

| Item | Total Graduate (1978-1987) | 1978 | 1980 | 1986 | 1987 |
|---|---|---|---|---|---|
| ***Institutions of Higher Learning*** | **2856.4** | **164.6** | **146.6** | **392.8** | **531.9** |
| *Engineering* | 888.3 | 56.5 | 44.2 | 119.2 | 156.1 |
| *Agriculture and Forestry* | 195.5 | 16.5 | 05.3 | 26.0 | 31.3 |
| *Medicine* | 270.7 | 27.5 | 17.7 | 27.9 | 32.1 |
| *Teacher Training* | 904.4 | 35.4 | 61.9 | 117.4 | 163.4 |
| *Liberal Arts* | 190.5 | 11.8 | 6.2 | 34.6 | 49.2 |
| *Science* | 173.4 | 12.8 | 8.4 | 21.6 | 24.7 |
| *Finance and Economics* | 158.8 | 1.6 | 1.3 | 32.7 | 53.9 |
| *Politics and Law* | 33.1 | 0.1 | 0.1 | 7.3 | 12.7 |
| *Physical Training* | 24.9 | 1.3 | 1.0 | 3.6 | 4.2 |
| *Art* | 16.8 | 1.1 | 0.6 | 2.5 | 4.3 |
| ***Specialized Secondary Schools*** | **4.127.6** | **232.0** | **410.3** | **496.2** | **577.8** |
| *Engineering* | 780.6 | 31.3 | 57.8 | 104.4 | 129.8 |
| *Agriculture and Forestry* | 351.4 | 27.1 | 32.3 | 40.0 | 42.7 |
| *Teacher Training* | 1683.2 | 112.7 | 209.1 | 175.0 | 189.0 |
| *Medicine* | 582.1 | 43.9 | 53.5 | 59.7 | 70.4 |
| *Finance and Economics* | 561.8 | 13.9 | 47.1 | 86.1 | 109.1 |
| *Politics and Law* | 95.3 | | | 20.6 | 23.4 |
| *Physical Training* | 12.6 | 0.2 | 1.0 | 1.7 | 2.0 |
| *Art* | 34.5 | 2.2 | 2.1 | 4.6 | 5.2 |
| *Other* | 26.1 | 0.7 | 7.4 | 4.1 | 6.2 |

## 9.4 *NUMBER OF ACHIEVEMENTS IN SCIENTIFIC AND TECHNICAL RESEARCH*

| Item | Total (1980-1987) | 1980 | 1985 | 1986 | 1987 |
|---|---|---|---|---|---|
| *Number of Major Achievements in Scientific and Technical Research* | 63447 | 2687 | 10476 | 14915 | 11800 |
| *Inventions Recognized and Rewarded by the State* | 1297 | 109 | 185 | 20 | 225 |

**Note:**
***The major achivements in scientific and technical research are evaluated and certified by government agencies at central or local levels.**

## 9.5 *ENROLLMENT IN ADULT EDUCATION BY LEVEL*
(1,000,000 persons)

| Classification | 1981 | 1982 | 1983 | 1984 | 1985 | 1986 | 1987 |
|---|---|---|---|---|---|---|---|
| ***Higher Education for Adults*** | **0.562** | **0.662** | **0.926** | **1.293** | **1.725** | **1.856** | **1.858** |
| *Radio and TV Universities* | 0.170 | 0.258 | 0.414 | 0.599 | 0.674 | 0.604 | 0.566 |
| *Staff and Worker, and Peasant Universities* | 0.105 | 0.144 | 0.174 | 0.193 | 0.261 | 0.341 | 0.339 |
| *Correspondence and Evening Universities* | 0.241 | 0.208 | 0.273 | 0.322 | 0.503 | 0.595 | 0.646 |
| *Management Cadres Colleges* | | | 0.002 | 0.015 | 0.040 | 0.056 | 0.056 |
| *Education Colleges* | 0.046 | 0.052 | 0.063 | 0.164 | 0.247 | 0.260 | 0.251 |
| ***Secondary Education for Adults*** | **8.207** | **10.804** | **9.748** | **5.987** | **5.470** | **8.066** | **10.470** |
| *Specialized Secondary Schools* | 4.441 | 4.454 | 3.303 | 0.827 | 1.347 | 1.512 | 1.682 |
| *Middle Schools* | 3.766 | 6.350 | 6.445 | 5.160 | 4.123 | 2.130 | 1.425 |
| *Schools for Technical Training* | | | | | | 4.424 | 7.363 |
| ***Primary Education for Adults*** | **9.736** | **7.566** | **8.172** | **9.322** | **8.338** | **12.614** | **13.518** |
| *Technical Courses* | | | | | | 6.760 | 9.303 |
| *Courses for Primary Education* | 3.523 | 3.606 | 2.885 | 3.234 | 3.148 | 2.321 | 1.737 |
| *Literacy Courses* | 6.213 | 3.960 | 5.287 | 6.088 | 5.190 | 3.533 | 2.478 |

## 9.6 *NUMBER OF BOOKS, MAGAZINES AND NEWSPAPERS PUBLISHED*

| Year | Books | | Magazines | | Newspapers | |
|---|---|---|---|---|---|---|
| | Total** | Total Copies Printed 1,000,000,000 | Total** | Total Copies Printed 1,000,000,000 | Total** | Total Copies Printed 1,000,000,000 |
| *1978* | 14987 | 3.774 | 930 | 0.762 | 186 | 12.776 |
| *1979* | 17212 | 4.072 | 1470 | 1.184 | 69 | 13.082 |
| *1980* | 21621 | 4.593 | 2191 | 1.125 | 188 | 14.042 |
| *1981* | 25601 | 5.578 | 2801 | 1.462 | 242 | 14.065 |
| *1982* | 31784 | 5.879 | 3100 | 1.514 | 277 | 14.001 |
| *1983* | 35700 | 5.804 | 3415 | 1.769 | 340 | 15.511 |
| *1984* | 40072 | 6.248 | 3907 | 2.182 | 458 | 18.076 |
| *1985* | 45603 | 6.673 | 4705 | 2.560 | 698 | 19.976 |
| *1986* | 51789 | 5.203 | 5248 | 2.402 | 791 | 19.390 |
| *1987* | 61346 | 6.250 | 5683 | 2.660 | 829 | 20.920 |

**Note:**
*** Figures for newspapers published in 1979-1987 are province, autonomous region and municipality level newspapers. Figures for other years also include prefectural-level newspapers.**

**** Total number refers to the number of different titles of books, different magazines and newspapers.**

## 9.7 NATURAL SCIENTIFIC AND TECHNICAL PERSONNEL IN STATE OWNED UNITS

| Item | 06/30/1978 | 1980 | 1984 | 1985 | 1986 | 1987 |
|---|---|---|---|---|---|---|
| ***Total Personnel (1,000,000)*** | 4.3451 | 5.2763 | 7.4662 | 7.8167 | 8.2531 | 8.8943 |
| *Engineering* | 1.5712 | 1.8622 | 3.1625 | 3.4041 | 3.5812 | 4.0122 |
| *Agriculture and Forestry* | 0.2942 | 0.3111 | 0.4346 | 0.4507 | 0.4652 | 0.4882 |
| *Health Sciences* | 1.2756 | 1.5302 | 2.0785 | 2.1610 | 2.2256 | 2.3250 |
| *Scientific Research* | 0.3103 | 0.3229 | 0.3348 | 0.3364 | 0.3656 | 0.3137 |
| *Teaching* | 0.8938 | 1.2499 | 1.4558 | 1.4645 | 1.6153 | 1.7552 |
| ***Percentage of Technical Personnel*** | 100.0 | 100.0 | 100.0 | 100.0 | 100.0 | 100.0 |
| *Engineering* | 36.1 | 35.3 | 42.4 | 43.6 | 43.4 | 45.1 |
| *Agriculture and Forestry* | 6.8 | 5.9 | 5.8 | 5.8 | 5.6 | 5.5 |
| *Health Sciences* | 29.4 | 29.0 | 27.8 | 27.6 | 27.0 | 26.2 |
| *Scientific Research* | 7.1 | 6.1 | 4.5 | 4.3 | 4.4 | 3.5 |
| *Teaching* | 20.6 | 32.7 | 19.5 | 18.7 | 19.6 | 19.7 |
| ***Number of Natural Scientific and Technical Personnel per 1,000 Staff and Workers in State Owned Units*** | 56.3 | 65.8 | 86.4 | 86.9 | 88.4 | 92.1 |

## 9.8 NUMBER OF CULTURAL, ART AND BROADCASTING UNITS

| Year | Film Projection Units (1,000) | Performing Troupes | Broadcasting Stations | Television Stations | Cultural Centers | Public Libraries | Museums |
|---|---|---|---|---|---|---|---|
| *1978* | 115.946 | 3150 | 93 | 32 | 2748 | 1256 | 349 |
| *1979* | 122.121 | 3482 | 99 | 38 | 2892 | 1651 | 344 |
| *1980* | 125.462 | 3533 | 106 | 38 | 2912 | 1732 | 365 |
| *1981* | 130.827 | 3483 | 114 | 42 | 2803 | 1787 | 383 |
| *1982* | 143.650 | 3460 | 118 | 47 | 2925 | 1889 | 409 |
| *1983* | 162.153 | 3444 | 122 | 52 | 2946 | 2038 | 467 |
| *1984* | 178.387 | 3397 | 167 | 93 | 3016 | 2217 | 618 |
| *1985* | 182.948 | 3317 | 213 | 202 | 2965 | 2344 | 711 |
| *1986* | 173.857 | 3195 | 278 | 292 | 2993 | 2406 | 777 |
| *1987* | 163.000 | 3084 | 386 | 365 | 2983 | 2437 | 826 |

## 9.9 NUMBER OF HEALTH CARE UNITS AND HOSPITAL BEDS

| Year | Number of Health Care Units (1,000) | | Hospital Beds (1,000,000) | | | Beds per 1,000 Population |
|---|---|---|---|---|---|---|
| | Total | Hospitals | Total | Rural Areas | Percentage of Beds in Rural Areas | |
| 1978 | 169.732 | 64.421 | 1.856 | 1.140 | 61.4 | 1.94 |
| 1979 | 176.793 | 65.009 | 1.932 | 1.192 | 61.7 | 1.99 |
| 1980 | 180.553 | 65.450 | 1.982 | 1.214 | 61.3 | 2.02 |
| 1981 | 190.126 | 65.911 | 2.017 | 1.214 | 60.2 | 2.02 |
| 1982 | 193.438 | 66.149 | 2.054 | 1.221 | 59.4 | 2.03 |
| 1983 | 196.017 | 66.662 | 2.110 | 1.240 | 58.8 | 2.06 |
| 1984 | 198.256 | 67.169 | 2.166 | 1.247 | 57.6 | 2.10 |
| 1985 | 200.866 | 59.614 | 2.229 | 1.267 | 56.8 | 2.13 |
| 1986 | 203.139 | 59.693 | 2.297 | 1.264 | 55.0 | 2.16 |
| 1987 | 204.960 | 60.429 | 2.405 | 1.277 | 53.1 | 2.23 |

## 9.10 NUMBER OF MEDICAL PERSONNEL IN HEALTH CARE UNITS

(1,000,000 persons)

| Year | Total | Medical Technical Personnel** | Number of Doctors* | | | | Senior Nurses and Nurses | Number of Doctors per 1,000 Population |
|---|---|---|---|---|---|---|---|---|
| | | | Total | Traditional Medicine | Western Medicine | Practitioners of Western Medicine*** | | |
| 1978 | 3.106 | 2.464 | 1.033 | 0.251 | 0.359 | 0.423 | 0.407 | 1.08 |
| 1979 | 3.344 | 2.642 | 1.088 | 0.258 | 0.395 | 0.435 | 0.421 | 1.12 |
| 1980 | 3.535 | 2.798 | 1.153 | 0.262 | 0.447 | 0.444 | 0.466 | 1.17 |
| 1981 | 3.796 | 3.011 | 1.244 | 0.290 | 0.516 | 0.436 | 0.525 | 1.25 |
| 1982 | 3.958 | 3.143 | 1.307 | 0.303 | 0.557 | 0.445 | 0.564 | 1.29 |
| 1983 | 4.090 | 3.253 | 1.353 | 0.313 | 0.588 | 0.450 | 0.596 | 1.32 |
| 1984 | 4.214 | 3.344 | 1.381 | 0.324 | 0.597 | 0.458 | 0.616 | 1.34 |
| 1985 | 4.313 | 3.411 | 1.413 | 0.336 | 0.602 | 0.473 | 0.637 | 1.35 |
| 1986 | 4.446 | 3.507 | 1.444 | 0.341 | 0.619 | 0.482 | 0.681 | 1.36 |
| 1987 | 4.564 | 3.609 | 1.482 | 0.346 | 0.645 | 0.487 | 0.718 | 1.38 |

**Note:**
* The term doctor does not include rural and urban part-time health personnel.

**Medical technical personnel includes doctors, senior and junior nurses and miscellaneous medical technical personnels.

*** Refer to those graduated from vocational medical schools.

## 9.11 MASS PARTICIPATION IN SPORTS AND NUMBER OF GRADED ATHLETES

| Year | Persons Reaching Standard of State Physical Training Program (1,000,000) | Sports Organized by Committees at County Level or Above | Graded Athletes* | | | |
|---|---|---|---|---|---|---|
| | | | Total | International Master Sportsmen | Master Sportsmen | First Grade |
| 1978 | 4.230 | 19019 | 729 | | 67 | 36 |
| 1979 | 6.253 | 20948 | 31347 | | 396 | 547 |
| 1980 | 8.557 | 22753 | 47214 | | 1147 | 804 |
| 1981 | 10.143 | 22983 | 58716 | | 910 | 1045 |
| 1982 | 8.448 | 26281 | 66761 | | 691 | 828 |
| 1983 | 25.293 | 25360 | 59596 | | 818 | 787 |
| 1984 | 33.423 | 30302 | 66503 | | 722 | 918 |
| 1985 | 43.622 | 26949 | 67392 | 115 | 852 | 1079 |
| 1986 | 50.129 | 25193 | 73399 | 107 | 1093 | 1390 |
| 1987 | 60.389 | 27531 | 75794 | 81 | 928 | 1220 |

Note:
* Number of graded athletes for each year is that of newly certified athletes.

## 9.12 WORLD SPORTS RECORDS BROKEN AND WORLD CHAMPIONSHIPS WON

| Year | World Records Broken or Surpassed | | | World Championships Won | | |
|---|---|---|---|---|---|---|
| | Number of Items | Number of Times | Number of Record Breaking Athletes | Number of Items | Number of Times | Number of World Championship Athletes |
| 1978 | 3 | 3 | 6 | 4 | 4 | 4 |
| 1979 | 13 | 26 | 32 | 12 | 12 | 20 |
| 1980 | 7 | 15 | 17 | 3 | 3 | 3 |
| 1981 | 8 | 18 | 15 | 25 | 25 | 53 |
| 1982 | 11 | 15 | 16 | 12 | 13 | 31 |
| 1983 | 13 | 18 | 25 | 37 | 39 | 50 |
| 1984 | 12 | 17 | 17 | 33 | 37 | 46 |
| 1985 | 5 | 9 | 6 | 42 | 46 | 70 |
| 1986 | 7 | 12 | 12 | 26 | 26 | 56 |
| 1987 | 22 | 41 | 24 | 64 | 69 | 72 |

# STATISTICAL COMMUNIQUE OF THE STATE STATISTICAL BUREAU OF THE PEOPLE'S REPUBLIC OF CHINA ON CHINA'S ECONOMIC AND SOCIAL DEVELOPMENT IN 1987

**February 23, 1988**

In 1987, under the leadership of the Chinese Communist Party and the State Council, the Chinese people of the various nationalities further implemented the policies of reform and opening up to the outside world and launched a movement to increase production and income and practice economy. As a result, production developed steadily, the domestic market remained brisk, economic exchanges with other countries expanded and new achievements were made in science, technology, culture, education, public health and sports undertakings. According to initial estimates, the annual GNP* reached 1,092 billion yuan, an increase of 9.4 percent over 1986; and the national income came to 915.3 billion yuan, 9.3 percent more than the previous year. The major problems in the economy were demand outstripping supply, shortages of some commodities, especially major non-staple products, and excessive price rises.

---

***Gross national product refers to value added of material and non-material production sectors, plus net income from abroad. It does not include value of goods and services used for intermediate consumption.**

Section I
## *AGRICULTURE*

Agriculture production went up. The gross agricultural output value in 1987 reached 444.7 billion yuan, up by 4.7 percent over 1986.

Of the output of major agricultural products, that of grain reached 402.41 million tons, 10.9 million tons more than the previous year, approaching the record high of 1984. The downward trend in the output of cotton and oil bearing crops over the previous two years was halted. The output of tea, cured tobacco, silkworm cocoons and fruit went up, while that of sugar bearing crops and jute and ambary hemp went down.

The output of major agricultural products was as follows:

| Item | 1987(tn) | % Increase Over 1986 |
|---|---|---|
| *Grain* | 402,410,000 | 2.8 |
| *Cotton* | 4,190,000 | 18.4 |
| *Oil Bearing Crops* | 15,250,000 | 3.5 |
| *Rapeseed* | 6,730,000 | 14.5 |
| *Sugarcane* | 46,850,000 | -6.7 |
| *Beetroot* | 7,970,000 | -3.9 |
| *Jute, Ambary Hemp* | 960,000 | -32.5 |
| *Cured Tobacco* | 1,640,000 | 19.5 |
| *Silkworm Cocoons* | 397,000 | 7.6 |
| *Tea* | 497,000 | 8.0 |
| *Fruit* | 15,510,000 | 15.1 |

As the people's motivation for managing and protecting forests was further enhanced, forest management was improved. New achievements were made in the afforestation of the plains and in the establishment of the "shelter forest belts" in northwest, north and northeast China, New progress was achieved in the establishment of timber and fast growing forest bases. But, wanton felling of trees occurred in some places, and forest fire prevention remained weak. A devastating forest fire swept through the Greater Xingan Mountains.

In animal husbandry, the number of grazing livestock increased steadily. The number of large animals and sheep at year's end was up over the beginning of the year. The output of milk, wool, poultry and eggs all rose. The number of live pigs, slaughtered pigs, the output of pork, all fell by varying degrees.

The output of major animal by products and livestock were as follows:

| Item | Unit | 1987(tn) | % Increase Over 1986 |
|---|---|---|---|
| ***Pork*** | (tn) | 17,800,000 | -0.9 |
| ***Beef and Mutton*** | (tn) | 1,410,000 | 16.9 |
| ***Milk*** | (tn) | 3,190,000 | 10.0 |
| ***Wool*** | (tn) | 208,000 | 12.4 |
| ***Slaughtered Pigs*** | head | 255,130,000 | -0.8 |
| ***Large Animals (year-end)*** | head | 126,330,000 | 6.2 |
| ***Pigs (year-end)*** | head | 326,400,000 | -3.2 |
| ***Sheep (year-end)*** | head | 178,350,000 | 7.3 |

Fishery kept up its high development rate. The output of aquatic products was 9.4 million tons, up by 14.1 percent over the previous year. The total catch of freshwater products increased by 17.8 percent, while that of marine products, by 11.5 percent.

Investment in agricultural production increased and the construction of farmland irrigation projects was strengthened, resulting in improved conditions for production. By the end of 1987, the aggregate power used by the country's farm machinery reached 247.2 billion watts, a 7.7 percent increase over 1986. There were 880,000 large and medium sized tractors, a 2 percent increase; 560,000 trucks, an increase of 11.8 percent; and the power used by irrigation and drainage equipment was 67.9 billion watts, a 12.4 percent rise. A total of 20.1 million tons of chemical fertilizers were applied throughout the year, up 4.1 percent from 1986. The total consumption of electricity in rural areas was 65.9 billion kwh, an increase of 12.3 percent. But the problem of not enough resources for further agricultural development remained and the output of major agricultural products still failed to meet the needs of the development of the national economy.

Reforms in the countryside continued to deepen and the rural economy developed fairly rapidly. The total product of society of rural areas** in 1987 was 904.1 billion yuan, 12.7 percent more than in the previous year. Of that, the output value of rural industry, construction, transportation and commerce went up by 21.8 percent, bringing its proportion in the total product of society of rural areas from 46.9 percent in 1986 to 50.8 percent to surpass the share of agricultural products for the first time.

**The total product of society of rural areas includes output value of agriculture and of collective and privately owned rural industries, construction, transport and commerce.

Section II
## *INDUSTRY*

In 1987 a balanced development of industrial production was witnessed. The gross industrial output value for the year reached 1,378 billion yuan, an increase of 16.5 percent. The figure is 1,182 billion yuan if the output value of village run industries is excluded, 14.6 percent over the previous year. The output value of the state owned sector went up by 11 percent, the collective sector by 25 percent, the private sector by 48 percent, and that of foreign funded enterprises (Sino foreign joint ventures, co operative enterprises and wholly foreign owned enterprises) by 98 percent.

Light and heavy industries developed in proportion. The 1987 gross output value of light industry was 656 billion yuan, 16.8 percent more than in 1986, while that of heavy industry reached 722 billion yuan, up 16 percent. The development of the power industry and the industries supplying agriculture was noticeably accelerated, while the basic industries, such as iron and steel, non ferrous metal and chemical raw materials, expanded steadily. In the consumer goods industry, the sectors using non-agricultural products as raw materials, especially makers of high quality, brand name consumer durables, grew most rapidly. However, the contradiction between the structure of industrial production and the demand still remained, and the supply of energy and some raw materials was sometimes inadequate.
The output of major industrial products was as follows:

| Item | Unit | 1987 | % Increase Over 1986 |
|---|---|---|---|
| ***Cotton Yarn*** | (tn) | 4,320,000 | 8.7 |
| ***Cloth*** | (m) | 16,700,000,000 | 1.4 |
| ***Woolen Fabrics*** | (m) | 260,000,000 | 3.0 |
| ***Machine Made Paper and Paperboards*** | (tn) | 10,080,000 | 1.0 |
| ***Sugar*** | (tn) | 5,110,000 | -2.6 |
| ***Cigarettes*** | cases | 28,810,000 | 11.0 |
| ***Bicycles*** | unit | 40,910,000 | 14.7 |
| ***Televisions Sets*** | unit | 19,380,000 | 32.8 |
| *Color Television Sets* | unit | 6,720,000 | 62.1 |
| ***Tape Recorders*** | unit | 18,630,000 | 6.0 |
| ***Cameras*** | unit | 2,390,000 | 18.2 |
| ***Washing Machines*** | unit | 9,920,000 | 11.0 |
| ***Refrigerators*** | unit | 3,980,000 | 76.8 |
| ***Coal*** | (tn) | 920,000,000 | 2.9 |
| ***Crude Oil*** | (tn) | 134,000,000 | 2.6 |
| ***Electricity*** | Kwh | 496,000,000,000 | 10.3 |
| *Hydroelectricity* | Kwh | 99,500,000,000 | 5.3 |
| ***Steel*** | (tn) | 56,020,000 | 7.3 |
| ***Rolled Steel*** | (tn) | 43,910,000 | 8.2 |

| | | | |
|---|---|---|---|
| ***Cement*** | (tn) | 180,000,000 | 9.2 |
| ***Timber*** | (cu. m) | 68,430,000 | 5.2 |
| ***Sulphuric Acid*** | (tn) | 9,620,000 | 26.1 |
| ***Soda Ash*** | (tn) | 2,370,000 | 10.5 |
| ***Chemical Fertilizers (100% effective)*** | (tn) | 17,030,000 | 22.0 |
| ***Chemical Insecticides*** | (tn) | 260,000 | 27.9 |
| ***Power Generating Equipments Capacity*** | Kw | 9,600,000 | 32.9 |
| ***Machines*** | unit | 146,000 | -10.6 |
| *Numerical Controlled and High Precision Machines* | unit | 4,267 | 6.4 |
| ***Motor Vehicles*** | unit | 472,000 | 27.7 |
| ***Tractors*** | unit | 40,000 | 39.2 |
| ***Locomotives*** | unit | 909 | 11.1 |
| ***Steel Ships (civilian use)*** | (tn) | 1,920,000 | 12.7 |

Industrial production combined well with marketing which meant better economic results. Profits and taxes of state owned industrial enterprises totalled 130.7 billion yuan, up 9.9 percent over th previous year. The turnover period for circulating funds was shortened from 109.1 days in 1986 to 105.4 days. Per capita productivity increased by 7.6 percent. Overall product quality was stable and in some cases improved. Energy consumption dropped and industrial departments saved over 20 million tons of energy, raising the energy saving rate to 4 percent. However, production costs increased and losses among loss suffering enterprises rose 4.6 percent. Overstock of a small number of slow selling products increased.

Industrial reforms focused on the management within enterprises. Statistics from 28 provinces, autonomous regions or municipalities showed that a full 68 percent of state owned industrial enterprises have introduced the director responsibility system, 82 percent of the large and medium sized state owned industrial enterprises have instituted various forms of the contract management responsibility system, and 46 percent of the small state owned industrial enterprises have either been given over to collective management or leased or contracted to individuals.
Horizontal economic associations developed, involving more and more areas in more diverse forms. By the end of 1987, the number of horizontally associated organizations made up chiefly of industrial enterprises above the county level had reached 6,780, involving investments of 12.57 billion yuan. The output value and profits of these associated organizations rose 38 percent and 38.1 percent respectively over the previous year, exceeding the overall industrial averages.

Section III

## *INVESTMENT IN FIXED ASSETS AND CONSTRUCTION*

The growth of investment in fixed assets began to slow down. In 1987, the country's investment in fixed assets totalled 351.8 billion yuan, 49.8 billion yuan or 16.5 percent more than in the previous year. This is lower than the 18.7 percent growth rate of 1986. Of the total, 226.2 billion yuan went to state owned enterprises, up 14.4 percent; 48 billion yuan to collective enterprises, up 22.4 percent; 77.6 billion yuan was made by individuals, up 19.6 percent. The overall scale of investment was still too great and there was an excessive number of outside plan or newly opened projects under construction.

In capital construction, the various localities implemented the principle of supporting productive or key projects and planned investment while curbing non-productive or unimportant projects and outside plan investment. The investment structure was also improved. In state owned units, investment in capital construction came to 132.4 billion yuan, up 12.6 percent over 1986. Of the total investment in capital construction, 30.9 billion yuan went into energy industry departments, up 25.5 percent over the year before; 20.4 billion yuan into transportation and communications departments, up 7.9 percent; 19.4 billion yuan into raw materials industries, up 31 percent; and 6.7 billion yuan into light and textile industries, up 15.9 percent. The proportion of investment in energy, raw materials and other basic industries rose from 33.5 percent in 1986 to 38 percent.

Construction of key projects was strengthened. An investment of 36.2 billion yuan was put into 206 key projects organized by the state and given reasonable construction schedules. All the major construction of key projects, including oilfields, power stations, coal mines, railway lines and ports, proceeded satisfactorily. A total of 102 large and medium sized capital construction projects and 193 single items attached to large and medium sized projects were completed and put into operation.

Newly added production capacities in capital construction projects include: 8.1 million kw of power generating capacity (the highest annual total since the founding of the People's Republic), 17.11 million tons of coal, 16.31 million tons of crude oil (including capacities added through oilfield renovation and transformation or other investments), 272 km of newly built railways put into operation, 433 km of double track railways, 204 km of electrified railways, 9.84

million tons of port cargo handling capacity, 300,000 tons of ethylene, 2.97 million tons of cement and 3.9 million weight boxes of plate glass.

Technological transformation of enterprises advanced steadily. State owned units invested 74.3 billion yuan in updating equipment, an increase of 19.9 percent. Of this amount, 25.6 billion yuan was invested in enlarging production capacity, an increase of 21.3 percent; 10.6 billion yuan in increasing the variety of products, up 8.3 percent; 4.2 billion yuan in improving product quality, up 7.3 percent; and 2.8 billion yuan in reducing energy consumption, up 24.4 percent. A total of 42,000 updating and transformation projects were completed and put into operation, adding 54.9 billion yuan to fixed assets.

Reforms in the building industry deepened. The contract responsibility system was applied to 117,000 projects undertaken by state owned construction enterprises, which covered a construction area of 160 million square meters, accounting for 83.9 percent of the country's total. Of these, 18,000 projects, covering a construction area of 36.93 million square meters and accounting for 19 percent of the total were contracted through bidding. Reform promoted the development of their business. In 1987, the gross output value of state owned construction enterprises increased by 7.7 percent over the year before and the per capita productivity rose 6.2 percent.

Geological surveys made marked progress. A total of 366 mineral bases with 64 kinds of minerals were discovered. Reserves of 31 billion tons of coal were verified, as were considerable gold, metal, non-metal ore and energy reserves. Deep sea geological surveying made headway, obtaining rich marine geological data and opening up a new sphere for geological studies. A total of 8.18 million meters of tunnelling was completed in 1987.

Section IV
## *TRANSPORT, POST AND TELECOMMUNICATIONS*

Further reforms were carried out in the transport industry, and technology was improved. The volume of passenger and freight transportation saw remarkable progress.

Transport services in 1987 were as follows:

| Item | 1987 | % Increase Over 1986 |
|---|---|---|
| ***Freight Traffic*** | 2,190.9 billion ton/km | 9.2 |
| *Railway* | 947.1 billion ton/km | 8.2 |
| *Highway* | 240.9 billion ton/km | 13.7 |
| *Waterway* | 939.7 billion ton/km | 9.8 |
| *Airway* | 660.0 million ton/km | 37.5 |
| *Oil/Gas Pipelines* | 62.5 billion ton/km | 2.1 |
| ***Passenger Traffic*** | 535.0 billion person/km | 11.0 |
| *Railway* | 284.3 billion person/km | 10.1 |
| *Highway* | 212.9 billion person/km | 11.6 |
| *Waterway* | 19.2 billion person/km | 5.5 |
| *Airway* | 18.6 billion person/km | 27.4 |
| *Cargo Volume Handled at Seaports* | 397.0 million tons | 3.3 |

Post and telecommunications made even greater progress. Business transactions by these departments totalled 3.89 billion yuan, an 18.3 percent increase from 1986. The number of telephone users in urban areas at the end of the year amounted to 2.93 million, an increase of 17 percent since 1986. Domestic express post service was available in 196 major cities.

The railway department was conducting the contract system for the second year, and results showed continued improvement. The average daily number of freight carriages loaded was over 70,000, an increase of 1.6 percent form 1986; average daily productivity of each locomotive was 821,000 ton km, an increase of 1.9 percent from 1986; income and profits from the railways increased 8.7 percent and 2.6 percent from 1986 respectively; labour productivity of railway workers rose 5.3 percent. Ships directly managed by the Ministry of Communications carried 50,100 ton/km of goods per ton of carrying capacity, an increase of 7.4 percent from 1986. Container transport developed comparatively quickly. In 1987, the volume of railway and sea container transport was 14.38 million tons, an increase of 27.7 percent from the previous year.

The number of passenger train carriages increased, and 40 new passenger trains were introduced. At the same time the railway service system was reformed and the number of sleepers for conductors and conductresses and freight carriages on express trains were reduced, allowing for more seat carriages and thereby more passengers.

The highway transport department did all it could to develop long distance passenger transport to lighten the burden on the railways. Interprovincial passenger transport routes in China totalled 2,500. The volume of truck freight transport rose 10.5 percent form the previous year. Still services in transport, post and telecommunications were inadequate, and the number of accidents was lamentably high.

Section V
## *DOMESTIC COMMERCE, SUPPLY AND MARKETING*

The domestic market remained brisk in 1987. The total value of retail sales was 582 billion yuan, a 17.6 percent increase form 1986. In real terms the increase was 9.6 percent if price changes were factored in. Of the total value of retail sales, the sales of means of agricultural production was 70.5 billion yuan, an increase of 22.4 percent, and of consumer goods 511.5 billion yuan, up 16.9 percent. Because some government offices, organizations and enterprises were extravagant and wasteful, using government funds for unnecessary tours and banquets, consumption of social groups rose dramatically, and the value of retail consumables bought by social groups reached 55.3 billion yuan, an increase of 19.7 percent from 1986.

The retail sales of all commodities increased since 1986. Retail sales in the state owned sector grew by 15.6 percent; in the collective sector, by 15.4 percent; in the public and private joint ownership sector, by 26.3 percent; and in the individual sector, by 24.2 percent. A 24 percent rise was registered in transactions between farmers and non-agricultural people.

Increases were registered in the sales of all consumer goods. Besides pork, vegetables, sugar and a few other food items which were somewhat in short supply, the availability of food was sufficient, and in fact 18.1 percent more than in 1986 with a value of 7.2 percent more in real terms. Sales of clothes increased by 13.9 percent and other necessities, 16.7 percent.

Structural reforms in commerce progressed. By the end of 1987, over 60 percent of the large and medium state owned commercial enterprises had implemented the contract management responsibility system; and 80 percent of the small state owned commercial enterprises were changed to collective management and collective ownership, or leased to individuals. Over 90 percent of the supply and marketing cooperatives carried out some form of responsibility system. The number of urban and rural fairs increased from 68,000 in 1986 to 69,000 in 1987. Their business volume in 1987 amounted to 110 billion yuan, a 21.3 percent rise over the preceding year. Commercial associations spread continuously, numbering 6,792 by the end of 1987, a 20.3 percent rise over 1986. Their turnover in 1987 was 15.5 billion yuan, and their profits was 650 million yuan.

There was an improvement in the economic results of state owned commerce and of supply and marketing cooperatives. Expenses for every 100 yuan in sales dropped 1.4 percent in 1987, and the profits therefrom grew by 6.4 percent. The time for the turnover of circulating funds was shortened from 220 days in 1986 to 203 days in 1987.

The scope of market regulation of capital goods continued to expand, entailing a drop in the proportion of materials distributed according to state plan. The proportion of rolled steel distributed under state plan dropped to 46.8 percent from 53.1 percent in 1986, of cement, to 15.6 percent from 16.2 percent in 1986, and of timber, to 26.2 percent from 30 percent in 1986. Sales of materials and capital goods departments totalled 156.7 billion yuan, a 32.5 percent rise over 1986, of which 26.8 billion yuan were handled by centers for trading in means of production, a growth of 69.6 percent.

Prices rose considerably. The general index of retail price for the year rose an average of 7.3 percent, while the price rise in December was 9.1 percent over December of 1986. The retail price index rose 9.1 percent in urban areas and 6.3 percent in the rural areas. Prices for foodstuffs went up 10.1 percent in 1987; the prices of meat, poultry and eggs rose 16.5 percent; of fresh vegetables, 17.7 percent; aquatic products, 17 percent. The prices of clothing articles rose by 3.5 percent (pure knitting wool rose 18.1 percent); household necessities, 6.1 percent; medicine and medical apparatus, 4.6 percent; fuels, 3.6 percent; and means of agricultural production, 7 percent.

The cost of living index for urban employees went up by an average of 8.8 percent in 1987 , and over 10 percent in a number of large and medium cities.

The purchasing prices of agricultural and sideline products rose by 12 percent in 1987.

The main problems in managing the market and prices were monopolized trades or enterprises arbitrarily raising the buying and selling prices of some scarce goods, and profiting a great deal by acting as business go-betweens. Not a few stores and peddlers raised prices in disguise by offering second rate goods as first rate, or giving customers less goods than they paid for.

Section VI

***FOREIGN TRADE AND TOURISM***

In 1987, China's exports grew, while the rise in imports was controlled. According to customs statistics, China's international trade in 1987 was worth 82.7 billion US dollars, 12 percent more than in 1986. Exports totalled 39.5 billion US dollars, up 27.8 percent; imports totalled 43.2 billion US dollars, up 0.7 percent. Last year's trade deficit was 3.7 billion US dollars, while in 1986 it was 12 billion.

Invisible trade continued to grow. In 1987, invisible earnings stood at 5.38 billion US dollars; with spending standing at 1.99 billion US dollars, earnings were 3.39 billion more than expenditure.

In 1987, China absorbed 7.57 billion worth of US dollars in foreign capital, 4.3 percent more than in 1986. This included 5.33 billion in loans, 6.4 percent more than in 1986; and 2.24 billion in direct foreign investment, more or less the same as in 1986.

Further advances were made in economic and technological cooperation with foreign countries. Contacts for overseas projects and labor services signed in 1987 were worth 1.74 billion US dollars, 28.3 percent more than in 1986. In 1987, China fulfilled contracts worth 1.11 billion, a 13.5 percent increase from 1986.

The economic strength of all the special economic zones expanded as did their export oriented character. Statistics from Shenzhen, Zhuhai, Shantou and Xiamen showed that the gross industrial output value of the four cities amounted to 11.2 billion yuan last year, an increase of 52 percent over 1986. Their total export value reached 2.74 billion US dollars, up 114 percent over 1986. More foreign investment went into productive projects. Last year saw over 100 foreign funded factories go into operation. A total of 510 million US dollars was utilized by the four cities last year, a drop of 16 percent from 1986.

The tourist industry also made great progress and in 1987 China attracted 26.9 million tourists and visitors from 169 countries and regions. This was 17.9 percent more than in 1986. Foreign exchange earned through tourism during the year amounted to 1.84 billion, up 20.3 percent over 1986.

Section VII

***SCIENCE, TECHNOLOGY, EDUCATION AND CULTURE***

Structural reform of science and technology continued to develop steadily, and science and technology played an increasingly significant role in China's development. In 1987, a total of 225 inventions and 807 technical advances won state prizes, and another 9,902 scientific and technological research results won ministry or local prizes. Of the technical advances, the mini-type neutron producer pile, the state laser wave length standard series, the ultraprecision axis series and the ultraprecision lathe were up to advanced world levels. To join the world's high tech development, nearly 800 research programs were approved last year, involving a total investment of 200 million yuan. The programs fell into seven fields including biological engineering, information technology, new energy resources and new materials. Nearly 4,000 contracts, as part of the key research projects for the Seventh Five Year Plan period (1986-1990), were decided upon by the end of last year. The State Fund for Natural Sciences, devoted mostly to supporting basic theoretical research and partly to applied science projects, subsidized 2,777 projects last year, to the tune of 130 million yuan. The "sparking program" which is meant to boost economic development in the rural areas, made good progress in 2,800 projects in the last two years. Scientific and technological activities among the masses were extensive. The technological market was blossoming and a total of 132,000 contracts were signed for transferring technology worth 3.35 billion yuan.

Patent work also developed last year. The year witnessed the Chinese Patent Office receiving 26,077 applications for patent rights and granting 6,811 patents, more than twice the 1986 figure.

Meteorological and marine departments made fairly prompt and accurate forecasts on weather and ocean conditions, providing a great deal of socially and economically helpful data.

To boost economic construction and access to resources, the cartographical department drew up 26,000 maps of varying scales. Altogether 223 maps were printed, with a total run of 82.22 million.

The ranks of scientists and technicians continued to expand. In 1987, a total of 8.68 million natural scientists and technicians were employed in state units, 430,000 more than in 1986. By the end of last year, China had 5,580 independent, state owned research and development institutes above county level, with a total work force of 1.05 million.

Education developed further under reform. Institutions of higher learning enrolled 40,000 postgraduates in 1987. There were 120,000 postgraduates studying, 28,000 of whom received postgraduate degrees. Universities and colleges enrolled 617,000 students last year, and the total student body was 1,959,000, up 4.2 percent from 1986. Last year, 532,000 students of higher education graduated. Institutions of adult higher education registered 498,000 new students in 1987, and thus had a total enrollment of 1,858,000.

The structure of secondary education headed for rationalization. The various vocational and technical schools had an enrollment of 5,169,000 students (including 1,046,000 at secondary technical training schools), 377,000 more than in 1986. The figure constituted 40 percent of the total number of students in senior middle schools. In 1987, there were 1,680,000 students at adult vocational schools and 7,360,000 at adult technical schools.

Elementary education was further strengthened. In 1987 there were 41.74 million students in junior middle schools and 128.36 million pupils in primary schools. The attendance rate of school age children increased from 96.4 percent in 1986 to 97.1 percent in 1987. Some 1,240 counties were recognized as popularizing elementary education. Much headway was made in special education for the physically disabled and mentally retarded.

The nation's culture flourished. In 1987, China produced 146 feature films and released 195 new full length films. The country boasts 163,000 cinemas and film projection teams, 3,089 performing art troupes, 2,980 cultural centers, 2,432 public libraries, 826 museums and 3,238 archives. In addition, there were also 385 broadcasting stations, 624 radio transmitting and relay stations, 365 television stations and 719 television transmitting and relay stations each with a capacity of more than 1,000 watts. Some 20.6 billion copies of national and provincial newspapers, 2.64 billion copies of magazines and 6.25 billion copies of books were published last year.

## Section VIII
## *PUBLIC HEALTH AND SPORTS*

Medicine and public health services improved. The number of hospital beds in China exceeded 2,365,000 at the end of 1987, 3 percent more than in 1986. Professional health workers numbered more than 3,609,000, up 2.9 percent since 1986. The total included 1,482,000 doctors (of whom 777,000 were senior doctors practising western medicine or traditional Chinese medicine), a 2.6 percent increase, and 718,000 nurses, up 5.4 percent. New successes were achieved in preventing and controlling various infectious and chronic diseases. The incidence of infectious diseases dropped 20 percent from 1986. Still city residents have difficulty getting an appointment, and the rural areas are short of doctors and medicines: 12.2 percent of villages have no clinics.

Notable achievements were made in sports. In 1987, China's athletes won 69 championships in world tournaments. They broke 22 world records on 41 occasions; they also broke 171 national records on 392 occasions. At the Sixth National Games they achieved record results.

## Section IX
## *PEOPLE'S LIVELIHOOD*

People's income in both cities and the countryside continued to increase. In 1987, the payroll of employees for the country totalled 186.6 billion yuan, an increase of 20.6 billion yuan, or 12.4 percent over 1986. The sample survey of urban wage earner households in 1987 showed an average annual per capita income of 916 yuan which can be used as living expenses, a 10.6 percent rise from 1986. If price rises are taken into account, the real per capita income rose 1.7 percent. The survey of farmers' families showed an average annual per capita net income of 463 yuan, 9.2 percent more than in 1986, or 5.3 percent in real terms. The total net income of farmers included 419 yuan from productive activities, an 11.7 percent increase. The income rise for urban and rural residents of different status was not balanced, and the real income of 21 percent of urban households dropped because of price rises. In the countryside, the average annual per capita net income of 8.2 percent of farm households was below 200 yuan.

Employment increased and reforms to the labor system progressed. In 1987, 3.49 million people found jobs in urban areas. The nation had 131.9 million employees by the end of 1987, 3.81 million more than the figure at the end of 1986. Of them, contract workers in state owned units numbered 7.26 million, up 2.02 million. Self employed workers in the cities and towns totalled 5.67 million, 840,000 more than in 1986.

Savings deposits increased greatly. By the end of 1987, individual bank savings amounted to 307.5 billion yuan, 83.8 billion or 37.5 percent more than the 1986 year end figure.

Housing for both urban and rural dwellers improved. In urban areas 180 million square meters of housing floor space were built in 1987, and 860 million in the countryside.

Social welfare services continued to improve. In 1987, the nation had 37,000 social welfare establishments, providing for 482,000 people throughout the country. Urban and rural collectives provided for 2.33 million elderly, disabled and orphans. Families who were suffering great hardships could find relief and support. Work and study opportunities and conditions for the disabled improved. Great progress was made in supporting poor and remote areas.

Section X
### *POPULATION*

China's natural growth rate increased slightly. According to a sample survey of one percent of the total population, the country's birth rate in 1987 was 21.04 per thousand, the death rate was 6.65 per thousand , yielding a natural growth rate of 14.39 per thousand, up from the 14.08 per thousand of 1986. The sample survey indicated that by the end of 1987 China had 1.08 billion people, 15 million more than the 1986 year end figure.

---

**Note: All figures given in this communique are preliminary statistics, and they do not include those for the Taiwan Province. Figures on gross national product, national income and gross output value are calculated at current prices, and the rates of growth are calculated at comparable prices.**

9 780275 932145

Printed by Libri Plureos GmbH in Hamburg,
Germany